Recent Progress in Materials Manufacturing

Recent Progress in Materials Manufacturing

Edited by **Keith Liverman**

WILLFORD PRESS

New York

Published by Willford Press,
118-35 Queens Blvd., Suite 400,
Forest Hills, NY 11375, USA
www.willfordpress.com

Recent Progress in Materials Manufacturing
Edited by Keith Liverman

International Standard Book Number: 978-1-68285-025-1 (Hardback)

The publisher's policy is to use permanent paper from mills that operate a sustainable forestry policy. Furthermore, the publisher ensures that the text paper and cover boards used have met acceptable environmental accreditation standards.

Trademark Notice: Registered trademark of products or corporate names are used only for explanation and identification without intent to infringe.

Printed in the United States of America.

Contents

Preface

Every book is initially just a concept; it takes months of research and hard work to give it the final shape in which the readers receive it. In its early stages, this book also went through rigorous reviewing. The notable contributions made by experts from across the globe were first molded into patterned chapters and then arranged in a sensibly sequential manner to bring out the best results.

The discipline of materials manufacturing has helped in developing better utilization techniques of raw materials, especially in industrial processes and technologies. This book on materials manufacturing unravels the recent studies in this field in a comprehensive manner. This book comprises of researches exploring various materials, their development, deployment and applications. It also highlights new models for advanced use of materials in manufacturing. It will be a resource guide for students and research scholars pursuing engineering and material sciences.

It has been my immense pleasure to be a part of this project and to contribute my years of learning in such a meaningful form. I would like to take this opportunity to thank all the people who have been associated with the completion of this book at any step.

Editor

Three dimensional modeling of complex heterogeneous materials via statistical microstructural descriptors

Yang Jiao[*] and Nikhilesh Chawla

* Correspondence:
yang.jiao.2@asu.edu
Materials Science and Engineering,
Arizona State University, Tempe, AZ
85287-6206, USA

Abstract

Heterogeneous materials have been widely used in many engineering applications. Achieving optimal material performance requires a quantitative knowledge of the complex material microstructure and structural evolution under external stimuli. Here, we present a framework to model material microstructure via statistical morphological descriptors, i.e., certain lower-order correlation functions associated with the material's phases. This allows one to reduce the large data sets for a complete specification of all of the local states in a microstructure to a handful of simple scalar functions that statistically capture the salient structural features of the material. Stochastic reconstruction techniques can then be employed to investigate the information content of the correlation functions, suggest superior and sensitive structural descriptors as well as generate realistic virtual 3D microstructures from the given limited structural information. The framework is employed to successfully model a variety of materials systems including an anisotropic aluminium alloy, a polycrystalline tin solder, the structural evolution in a binary lead-tin alloy when aged, and a model structure of hard-sphere packing. Our framework also has ramifications in the development of integrated computational material design schemes and 4D materials modeling techniques.

Keywords: Heterogeneous materials; 3D microstructure modeling; Statistical structural descriptors; Stochastic reconstruction

Background

Heterogeneous materials including metallic alloys, ceramics, composites and granular media have many important engineering applications. Such materials possess complex microstructures spanning a wide spectrum of length scales, which determine their macroscopic properties and performance [1-4]. The growing demands in global sustainability, national security, and renewable energy have raised great challenges in the development of multifunctional materials that can achieve optimal performance under extreme conditions [5,6]. The success of such a mission strongly relies on our ability to characterize and modify material properties and behaviors under a myriad of external stimuli. Thus, an intrinsic understanding and knowledge of complex microstructures and how they evolve under various conditions is extremely important.

Advances in experimental methods, analytical techniques, and computational approaches, have now enabled the development of three dimensional (3D) analyses [7].

Figure 1 illustrates the characterization of the microstructure of a Sn-3.5Ag eutectic alloy, a candidate lead-free solder for electronic packaging, by tomographic microscopy across a wide range of length scales. The study of 3D microstructures under an external stimulus (e.g., stress, temperature, environment) as a function of time (4D) is particularly exciting. Examples include an understanding of time-dependent deformation structures, phase transformations, compositional evolution, magnetic domains, etc. Furthermore, advances in 3D and 4D computational tools and methods have enabled the analysis of large experimental data sets, as well as simulation and prediction of material behavior [8-10], which we will briefly discuss in Methods.

X-ray tomography is an extremely attractive, non-destructive technique for characterizing microstructures in 3D and 4D. The use of high brilliance and partially coherent synchrotron light allows one to image multi-component materials from the sub-micrometer to nanometer range. X-ray tomography can be conducted in imaging modes based on absorption or phase contrast. The technique can also be used using lab-scale systems. To construct an accurate 3D microstructure model, 2D projections are obtained at small angular increments. Given a sufficiently large number of such 2D projections, tomographic reconstruction techniques such as the filtered-back-projection algorithm [11] can be employed to generate a grayscale image of the material microstructure. Further segmentation and thresholding analysis are used to resolve details of individual material phases and produce accurate digital representations of the 3D microstructure. These 3D or 4D data sets can be used to quantify the microstructure, and/or can be used as an input for microstructure-based modeling. Thus, tomography is an excellent technique that eliminates destructive cross-sectioning, and allows for superior resolution and image quality with minimal sample preparation [12-19].

Experimentally obtained microstructural data are usually represented as a large array whose entries indicate the local states of the microstructure (e.g., the phase that a specific voxel belongs to). Although morphological details of a specific material can be contained in its microstructrual array, such information would not all be useful for macroscopic effective property analysis, which requires averaging over a sufficiently large number of material samples to yield meaningful and robust statistics. In the case where the entire course of microstructural evolution is of interest (e.g., during

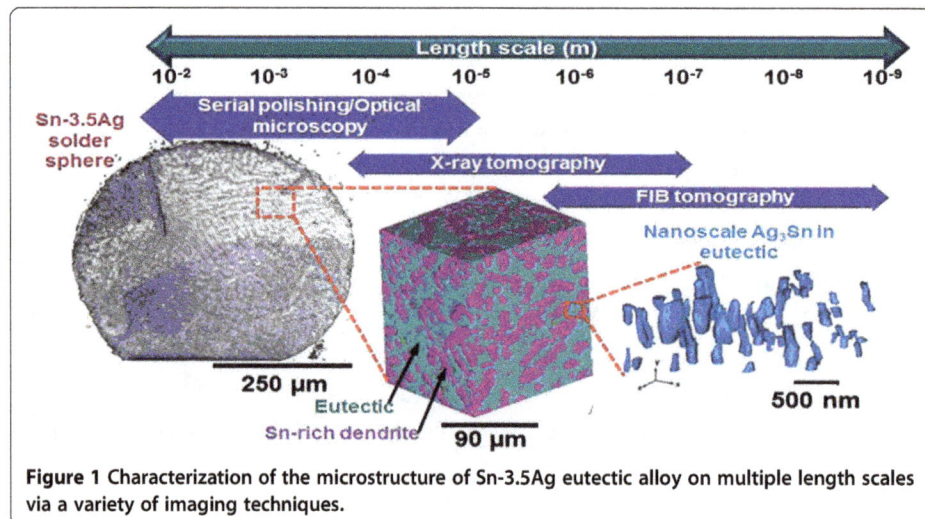

Figure 1 Characterization of the microstructure of Sn-3.5Ag eutectic alloy on multiple length scales via a variety of imaging techniques.

the solidification or coarsening processes), the resulting 4D structural data sets (three spatial dimensions plus one temporal dimension) are usually extremely large. In certain 4D cases, such as the progression of a fatigue crack over time, most of the material containing the crack remains unaltered and the major structural changes only occur in the vicinity of the crack tip [19]. Therefore, representing this process using a large number of reconstructed 3D microstructures at successive time points during the progression of the crack is less efficient and results in a huge amount of redundant structural information. Therefore, it is highly desirable to devise a robust framework that enables one to represent, characterize and model the complex microstructure of a heterogeneous material in 3D and 4D in a much more efficient manner.

Furthermore, for certain systems such as a polycrystalline material, it is still extremely difficult to directly obtain the 3D microstructure containing the information on grain misorientations using current imaging techniques. A common practice is to use serial sectioning to reconstruct the 3D material from 2D electron-back-scattered-diffraction (EBSD) images of the material's surfaces [20,21]. This technique, which involves extensive surface polishing, image smoothing and image alignment, is extremely time consuming and tedious. Thus, developing efficient statistical morphological descriptors that capture the key structural features of the materials from a handful of 2D images and that enable one to reconstruct accurate 3D virtual microstructure models from such limited information will significantly improve the efficiency in characterizing polycrystalline materials.

In this article, we present a framework to model the microstructure of complex heterogeneous materials via statistical morphological descriptors, i.e., certain lower-order correlation functions associated with the material's phases [3]. Such correlation functions are mainly derived from homogenization theories that quantitatively connect the macroscopic properties of a heterogeneous material to its complex microstructure [2-4]. Representing a microstructure with the combination of a selected set of correlation functions allows one to reduce the large data sets for a complete specification of all of the local states in a microstructure to a handful of simple scalar functions that statistically capture the salient structural features of the material [22-24]. Stochastic reconstruction techniques such as the simulated annealing procedure developed by Yeong and Torquato [25,26] can then be employed to investigate the information content of the correlation functions, suggest superior and sensitive structural descriptors as well as generate realistic virtual 3D microstructures from the given limited structural information. Material modeling based on correlation functions has enabled the development of data-driven material design schemes [27,28] and 4D materials modeling techniques [29].

The rest of the paper is organized as follows: In Statistical microstructural descriptors, we provide definitions of the statistical morphological descriptors (i.e., correlation functions) employed to model complex heterogeneous materials. In Methods, we present the Yeong-Torquato reconstruction procedure to generate virtual 3D microstructure models from a prescribed set of correlation functions. In Results, we apply our general framework and the reconstruction technique to model a variety of material systems, including an anisotropic aluminum alloy, a polycrystalline tin solder, as well as the 4D coarsening process of a lead-tin binary alloy. Concluding remarks are provided in Discussion and conclusion.

Statistical microstructural descriptors

In general, the microstructure of a heterogeneous material can be uniquely determined by specifying the indicator functions associated with all of the individual phases of the material [3], i.e.

$$I^{(i)}(\mathbf{x}) = \begin{cases} 1 & \mathbf{x} \text{ in phase } i \\ 0 & \text{otherwise} \end{cases} \tag{1}$$

where $i = 1, ..., q$ and q is the total number of phases. In practice, a digitized version of the indicator function, i.e., a 3D array whose entries specifying the phases that the voxels belong to, is used to fully characterize a material microstructure. There exist several approximation schemes for $I^{(i)}(\mathbf{x})$ (i..e, the material microstructure), including the Gaussian random field (GRF) models [30] and reduced-order representation [31]. The key idea involved in these approximation schemes is to reduce the large number of degrees of freedom required to specify unnecessary structural details to a small set of parameters or data points that statistically characterize the material microstructure for purpose of modeling and visualization. However, the physical significance associated with these parameters is usually not obvious. On the other hand, as we will show in the ensuing sections, the parameters involved in the correlation functions such as the "correlation length" often have clear physical meaning and directly correspond to a salient structural feature of the material.

Two-point correlation function

Given the indictor function $I^{(i)}(\mathbf{x})$, the volume fraction of phase i is then given by

$$\phi_i = <I^{(i)}(\mathbf{x})> \tag{2}$$

where <> denotes the ensemble average over many independent material samples or volume average over a single large sample if it is spatially "ergodic" [3]. The two-point correlation function $S_2^{(ij)}(\mathbf{x}_1, \mathbf{x}_2)$ associated with phases i and j is defined as

$$S_2^{(ij)}(\mathbf{x}_1, \mathbf{x}_2) = <I^{(i)}(\mathbf{x}_1)I^{(j)}(\mathbf{x}_2)> \tag{3}$$

which also gives the probability that two randomly selected points $\mathbf{x}_1$ and $\mathbf{x}_2$ fall into phase i and j respectively (see Figure 2). For a material with q distinct phases, there are totally $q*q$ different S_2. However, it has been shown that only q of them are independent [3,32] and the remaining $q*(q-1)$ functions can be explicitly expressed in terms of the q independent ones.

For statistically homogeneous materials as the examples considered here, there is no preferred center in the microstructure. Therefore, the associated S_2 depends only on the relative vector displacement between the two points [3], i.e.,

$$S_2(\mathbf{x}_1, \mathbf{x}_2) = S_2(\mathbf{x}_2 - \mathbf{x}_1) = S_2(\mathbf{r}) \tag{4}$$

where $r = x_2 - x_1$. At $r = 0$ the auto-correlation function gives the probability that a randomly selected point falls into the phase of interest, i.e., the volume fraction of the associated phase. On the other hand, the cross correlation function is zero at $r = 0$, since the probability of finding a single point falling into two different phases is zero. At large r values, the probabilities of finding the two points in the phases of interest are independent of one another, leading to ϕ_i^2 for the auto-correlation

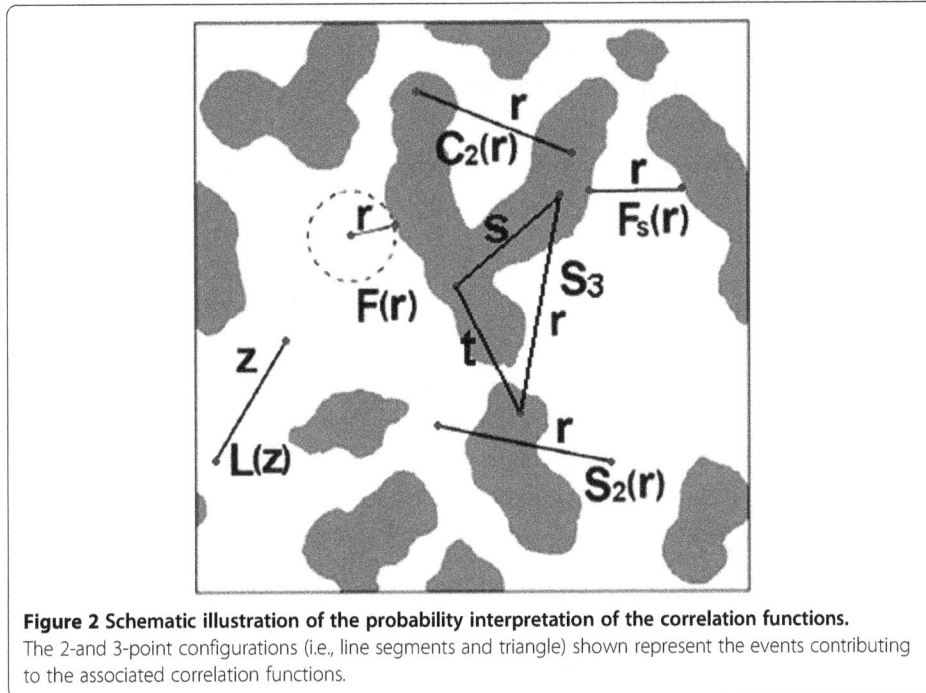

Figure 2 Schematic illustration of the probability interpretation of the correlation functions.
The 2-and 3-point configurations (i.e., line segments and triangle) shown represent the events contributing
to the associated correlation functions.

functions and $\phi_i\phi_j$ for the cross correlation function. For a statistically isotropic material, S_2 depends only on the scalar distance between a pair of points.

We note that the general n-point correlation function S_n which gives the probability of finding a particular n-point configuration in specific phases can be defined in a similar manner as S_2; see Eq. (2). It has been shown that the effective properties of a heterogeneous material can be explicitly expressed as series expansions involving certain integrals of S_n. Interested readers are referred to Ref. [3] for detailed discussions of S_n and their properties.

Lineal-path function

The lineal-path function $L^{(i)}(r)$ gives probability that a randomly selected line segment of length $r = |\, r\, |$ along the direction of vector r entirely falls into phase i (see Figure 2) [33,34]. At $r = 0$, $L^{(i)}(0)$ reduces to the probability of finding a point in phase i and thus, $L^{(i)}(0) = \phi_i$. In materials that do not contain system-spanning clusters, the chance of finding a line segment with very large length entirely falling into any phases is vanishingly small. Accordingly, for large r values $L^{(i)}$ decays to zero rapidly in such materials. The lineal-path function contains partial topological connectedness information of the material's phases, i.e., that along a lineal-path. Generally, the lineal-path function underestimates the degree of clustering in the system (e.g., two points belonging to the same cluster but not along a specific lineal path will not contribute to L).

Two-point cluster function

The two-point cluster correlation function $C_2^{(i)}(\mathbf{x}_1, \mathbf{x}_2)$ gives the probability that two randomly selected points $\mathbf{x}_1$ and $\mathbf{x}_2$ fall into the same cluster of phase i (see Figure 2) [35,36]. For statistically homogeneous materials, C_2 depends only on the relative vector displacement between the two points, i.e., $C_2(\mathbf{x}_1, \mathbf{x}_2) = C_2(\mathbf{r})$. In contrast to the lineal-path function, C_2 contains complete clustering information of the phases, which has been shown to have dramatic effects on the material's physical properties [3]. Moreover, unlike S_2 and L, the cluster functions generally cannot be obtained from lower-dimensional cuts (e.g., 2D slices) of a 3D microstructure, which may not contain correct connectedness information of the actual 3D system.

It has been shown that C_2 is related to S_2 via the following equation [35]

$$S_2^{(ii)}(\mathbf{r}) = C_2^{(i)}(\mathbf{r}) + D_2^{(i)}(\mathbf{r})$$ (5)

where $D_2^{(i)}(\mathbf{r})$ measures the probability that two points separated by r fall into different clusters of the phase of interest. In other words, C_2 is the connectedness contribution to the standard two-point correlation function S_2. For microstructures with well-defined inclusion, $C_2(r)$ of the inclusions is a short-ranged function that rapidly decays to zero as r approaches the largest linear size of the inclusions. We note that although C_2 is a "two-point" quantity, it has been shown to embody higher-order structural information which makes it a highly sensitive statistical descriptor over and above S_2 [24].

Other correlation functions

Other types of correlation functions include the surface functions and pore-size function. The surface correlation functions F_s [3,37] are associated with the probability of finding an n-point configuration with a subset of points falling on to the surface, while the others lie in the bulk of the phase of interest. Specifically, the surface-surface correlation function $F_{ss}(r)$ can be considered as the probability of finding two points separated by r, which both fall into a dilated region associated with the surface of one phase in the limit of vanishingly small dilation. The surface-void correlation function $F_{sv}(r)$ is the probability of finding two points separated by r, one in the phase of interest ("void") and the other in a dilated region associated with the surface of one phase in the limit of vanishingly small dilation. The pore-size function F [38,39] provides statistics of the size of the largest spherical region centered at a randomly selected point in the phase of interest that entirely lie in this phase. Figure 2 schematically illustrates the probability interpretation of the aforementioned statistical microstructural descriptors by depicting the events that contribute to the correlation functions.

Modeling microstructure via correlation functions

It is rigorously shown that a heterogeneous microstructure can be uniquely determined given the complete set of n-point correlation functions S_n. In practice, such a complete set is generally not available and is not suitable for microstructure modeling due the complexity involved in computing these quantities from experimental images. As we mentioned above, the non-traditional functions such as the lineal-path function L and the cluster function C_2 contain topologically connectedness information that is only embodied in higher-order S_n. Therefore, we expect that a carefully selected set of aforementioned lower-order correlation functions (e.g., S_2, C_2, L, F_{ss}, F_{sv}, F, etc.), which capture different structural features such as connectivity and surface of the material system, could provide a statistically accurate characterization of the microstructure. In other words, a microstructure M can be modeled and represented as a set of selected lower-order correlation functions, i.e.,

$$M = \{S_2, C_2, L, F_{ss}, F_{sv}, F, ...\}$$ (6)

Methods

A variety of reconstruction schemes have been developed that enable one to generate material microstructure models from limited structural information. Examples of such schemes include the Gaussian random field method [30], phase recovery method [40], multi-point reconstruction method [41], and raster-path method [42]. The Gaussian random field method [30] was originally devised to reconstruct realizations of statistically homogeneous and isotropic random media from the associated two-point correlation functions. The phase recovery method enables one to take into account the full vector information contained in the two-point statistics associated with the material, and thus allows the reconstructions of complex anisotropic microstructure and polycrystalline materials [40]. The multi-point reconstruction method was originally developed for the reconstruction of porous geomaterials from 2D images [41]. Instead of using two-point statistics associated with the entire 2D microstructure, this method incorporates all n-point statistics within a smaller window containing a portion of the microstructure. The recently developed raster-path method allows one to employ the multi-point statistics in a much more efficient way [42]. Specifically, instead of extracting the statistics from the 2D microstructure, a cross-correlation function is introduced to directly compare a reconstructed portion of the material to the target 2D image.

All of the aforementioned reconstruction methods have been successfully applied to model certain classes of materials. However, they all require specific correlation functions or image data as input information. Another widely-used reconstruction method is the stochastic optimization procedure devised by Yeong and Torquato [25,26]. The Yeong-Torquato (YT) procedure enables one to incorporate an arbitrary number of correlation functions of any type into the reconstructions. Specifically, a trial microstructure is evolved using simulated annealing procedure such that the set of correlation functions sampled from the trial microstructure match the set of target statistics up to a prescribed small tolerance (see below for algorithmic details). Its flexibility of incorporating arbitrary correlation functions into the reconstruction makes the YT procedure an ideal protocol to examine the information content of different statistical descriptors associated with a material system and select the most sensitive descriptors for microstructure modeling. Therefore, we will employ the YT procedure in our study. However, due to its stochastic nature a large number of intermediate trial microstructures need to be generated and analyzed, which makes it computationally intensive. Several different implementations of the Y-T procedure have been devised to improve it efficiency [22-24,43,44].

Yeong-Torquato reconstruction procedure

We use the Yeong-Torquato (YT) reconstruction procedure [25,26] to generate virtual 3D microstructures from a specific set of correlation functions discussed in the previous Statistical microstructural descriptors. In the YT procedure, the reconstruction problem is formulated as an "energy" minimization problem, with the energy functional E defined as follows

$$E = \sum_{\alpha} \sum_{r} \left[f^{\alpha}(r) - \overline{f^{\alpha}}(r) \right]^2 \tag{7}$$

where $\overline{f^{\alpha}}(r)$ is a target correlation function of type α and $f^{\alpha}(r)$ is the corresponding function associated with a trial microstructure. The simulated annealing method [45] is usually employed to solve the aforementioned minimization problem. Specifically, starting from an initial trial microstructure (i.e., *old* microstructure) which contains a fixed number of voxels for each phase consistent with the volume fraction of that phase, two randomly selected voxels associated with different phases are exchanged to generate a *new* trial microstructure. Relevant correlation functions are sampled from the new trial microstructure and the associated energy is evaluated, which determines whether the new trial microstructure should be accepted or not via the probability:

$$p_{acc}(old \rightarrow new) = \min\left\{1, \exp\left(\frac{E_{old}}{T}\right) \middle/ \exp\left(\frac{E_{new}}{T}\right)\right\} \tag{8}$$

where T is a virtual temperature that is chosen to be initially high and slowly decreases according to a cooling schedule [23]. Initially, T is chosen to be high in order to allow a sufficiently large number of "up-hill" (energy increasing) trial microstructures. As T gradually decreases, the "up-hill" moves become less favorable. This significantly decreases the chances that the final microstructure gets stuck in a shallow local energy minimum, as illustrated in Figure 3. In practice, it is usually extremely difficult for the system to converge to the global minimum. Thus, the annealing process is considered complete if E is smaller than a prescribed tolerance, which we choose to be 10^{-10} here.

Efficient sampling method for correlation functions

The probabilistic interpretations of the correlation functions discussed in Statistical microstructural descriptors enable us to develop a general sampling method for reconstruction of statistically homogeneous and isotropic materials based on the "lattice-gas" formalism [22-24]. In this formalism, pixels with different values (occupying the lattice sites) correspond to distinct local states and pixels with the same value are considered to be "molecules" of the same "gas" species. The correlation functions of interest can be obtained by binning the separation distances between the selected pairs of molecules from particular species.

In the case of S_2, all molecules are of the same species. We denote the number of lattice-site separation distances of length r by $N_S(r)$ and the number of molecule-pair separation distances of length r by $N_P(r)$. Thus, the fraction of pair distances with both ends occupied by the phase of interest, i.e., the two-point correlation function, is given by $S_2(r) = N_P(r)/N_S(r)$. To obtain C_2, one needs to partition the molecules into different subsets Γ_i (species) such that any two molecules of the same species are connected by a path composed of the same kind of molecules, i.e., molecules that form a cluster, which

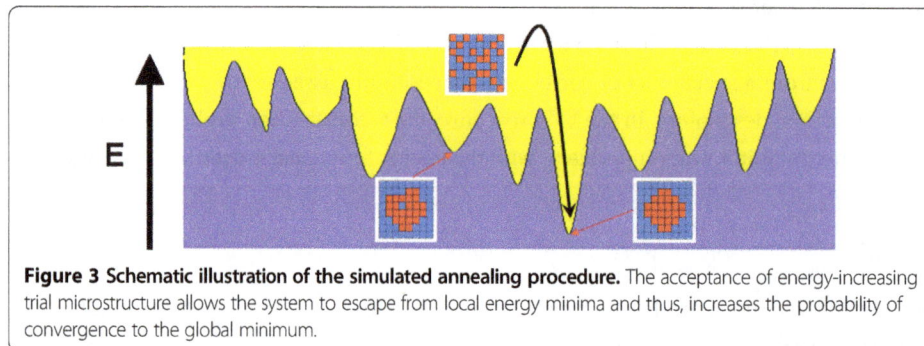

Figure 3 Schematic illustration of the simulated annealing procedure. The acceptance of energy-increasing trial microstructure allows the system to escape from local energy minima and thus, increases the probability of convergence to the global minimum.

is identified using the "burning" algorithm [24]. The number of pair distances of length r between the molecules within the same subset Γ_i is denoted by $N_P^i(r)$. The two-point cluster function is then given by $C_2(r) = i\ N_P^i(r)/N_S(r)$. The calculation of F_{SS} and F_{SV} requires partitioning the molecules into two subsets: the surface set K_S containing only the molecules on the surfaces of the clusters and the volume set K_V containing the rest. In a digitized medium, the interface necessarily has a small but finite thickness determined by the pixel size. Thus, the surface–surface and surface–void correlation functions can be regarded as probabilities that are given by $F_{SS} = N_{SS}(r)/N_S(r)$ and $F_{SV} = N_{SV}(r)/N_S(r)$, respectively; where $N_{SS}(r)$ gives the number of distances between two surface molecules with length r and N_{SV} is the counterpart for pairs with one molecule on the surface and the other inside the cluster.

The lineal path function L can be obtained by computing the lengths of all digitized line segments (chords) composed of pixels of the phase of interest, and for each chord incrementing the counters associated with the distances equal to and less than that chord length [24]. The chord-length density function p can then be easily obtained by binning the chord lengths that are used to compute L. The pore-size function F can be computed by finding the minimal separation distances of pixels within the phase of interest to those at the two-phase interfaces. The minimal distances are then binned to obtain a probability distribution function, the complementary cumulative distribution function of which is the pore-size function F [24].

Efficient method for re-computing correlation functions for trial microstructures

Generally, several hundred thousand trials need to be made to achieve such a small tolerance. Therefore, efficient sampling methods [22-24] are used that enable one to rapidly obtain the prescribed correlation functions of a new microstructure by updating the corresponding functions associated with the old microstructure, instead of completely re-computing the functions. Specifically, a distance matrix D that stores the separation distances of all "molecule" pairs is established when the system is initialized and the molecules are partitioned into different "species" depending on their positions, as discussed in the main article. The quantities N_P, N_P^i, N_{SS} and N_{SV} (all defined in the previous subsection) can be obtained by binning the separation distances of selected pairs of molecules from particular species. Recall that the molecules are pixels associated with different values indicating the species, i.e., the clusters or the surface/volume set they belong to.

In the reconstruction procedure, a trial microstructure is generated by moving a randomly selected pixel of the phase of interest to an unoccupied site. This results in changes of the separation distances between the moved the pixel and all of the other pixels, and causes two kinds of possible species events. The first kind is a "cluster" event, which involves breaking and combining clusters. For example, if the selected pixel happens to be a "bridge" connecting several sub-clusters, removing the bridge will make the original single cluster break into smaller pieces, i.e., new species (clusters) are generated. Similarly, the reverse of the above process can occur, i.e., a randomly selected pixel moved to a position where it connects several small clusters to form a larger cluster, which leads to combination of clusters and annihilation of species. The other kind of species event is the transition of pixels between surface and volume sets. If a pixel originally on the surface is removed, certain volume pixels (i.e., those inside the phase of interest) will now constitute the new surface and vice versa. The selected pixel itself could also undergo such a transition, depending on its original and new positions, e.g., a volume pixel originally inside the phase of interest could be moved to the interface to become a surface pixel.

The contributions of the number of pair distances to N_P, N_P^i, N_{SS} and N_{SV} from the pixels in old microstructure involved in the species events are computed and subtracted accordingly. The new contributions can be obtained by binning the separation distances of pixel pairs in the new microstructure involved in the species event, which are then added to the corresponding quantities N_P, N_P^i, N_{SS} and N_{SV}. This method only requires operations on a small number of pixels, including retrieving and binning their separation distances and updating the species sets (i.e., the clusters and surface/volume set). The use of the distance matrix D speeds up the operations involving distances. However, for very large systems (e.g., those including millions of pixels), storing D requires very a large amount of computer memory. An alternative is to re-compute the separation distances of the pixel pairs involved in the species events for every trial microstructure, instead of explicitly storing all the distances in D. This may slightly slow down the reconstruction process but make it easy to handle very large systems. Correlation functions of the new microstructure can then be obtained from the updated N_P, N_P^i, N_{SS} and N_{SV} (i.e., dividing those quantities by N_S). Importantly, the complexity of the algorithm is linear in the total number of pixels (molecules) within the system.

Results

In this section, we apply the aforementioned statistical descriptors and stochastic reconstruction techniques to model several material systems including an anisotropic aluminum alloy, a polycrystalline tin solder and the structural evolution in an isothermally-aged binary lead-tin alloy. Moreover, we examine the information content of different correlation functions by studying a model material microstructure consisting of equal-sized hard spheres in a matrix.

Inclusion distribution anisotropy in rolled aluminum alloy

Aluminum (Al) alloys have been widely used in many engineering structures especially in automotive and aircraft applications due to their unique high strength-to-weight ratio and corrosion resistance. An aluminum alloy almost always has secondary phase inclusions and particles with impurities that are present in the microstructure. Due to rolling of the alloy, the inclusions usually possess an anisotropic distribution, with the dimension along the rolling direction being significantly elongated. This in turn results in an overall anisotropic alloy microstructure.

In order to model an alloy microstructure with anisotropic Fe-rich and Si-rich inclusions in Al matrix, we will use correlation functions along three orthogonal directions, i.e., the longitudinal or rolling (L), transverse (T) and short transverse (S) directions [46]. Specifically, we denote the directional two-point correlation function, two-point cluster function and lineal-path function of type α along direction β respectively by $S_2^{\alpha,\beta}(r)$, $C_2^{\alpha,\beta}(r)$ and $L^{\alpha,\beta}(r)$ where $\alpha = \{Fe, Si, Fe\text{-}Si\}$ and $\beta = \{L, T, S\}$.

Accordingly, in order to obtain these functions from a digitized microstructure, only the "molecules" along specific directions are considered (see Results).

The directional two-point correlation function of the inclusion phase α can be approximated with the following function:

$$S_2^{\alpha,\beta}(r) = \left\{ \phi_\alpha \left[A_\beta^\alpha \exp\left(B_\beta^\alpha \cdot r\right) + \left(1 - A_\beta^\alpha\right) V_{\text{int}}\left(r; \bar{D}_\beta^\alpha\right) \right] + \phi_\alpha^2 \left[1 - V_{\text{int}}\left(r; \bar{D}_\beta^\alpha\right)\right] \right\} \Theta\left(\bar{D}_\beta^\alpha - r\right)$$
$$+ \phi_\alpha^2 \Theta\left(r - \bar{D}_\beta^\alpha\right)$$

$$(9)$$

where ϕ_α is the volume fraction, A_β^α, B_β^α and $\bar{D}_\beta^\alpha$ are parameters to be determined. The approximation (9) includes two "basis" functions: the Debye random medium function $\exp(Br)$ [22,47] and $V_{\text{int}}(r; D)$ [3], which is the scaled intersection volume of two isotropic inclusions with linear size D separated by r:

$$V_{\text{int}}(r; D) = 1 - \frac{3r}{2D} + \frac{r^3}{2D^3}$$

$$(10)$$

and $\Theta(x)$ is the Heaviside function, i.e.,

$$\Theta(x) = \begin{cases} 1 & x \geq 0 \\ 0 & x < 0 \end{cases}$$

$$(11)$$

In other words, one can consider the alloy microstructure as a mixture of "clusters of all shape sizes" (features of a Debye random medium) and well defined inclusions (characterized by V_{int}). For the anisotropic inclusions, we consider that each direction possesses distinct combination parameter A_β^α, correlation parameter B_β^α and effective size $\bar{D}_\beta^\alpha$, which is the effective linear size or "length" of the inclusions along that direction. The values of these parameters are determined such that the approximated functions best match the actual autocorrelation function (i.e., the squared difference between two functions is minimized). In Tables 1 and 2, we provide the values of the parameters for the Fe-rich and Si-rich inclusions in the alloy, respectively. Note that we do not model the cross-correlation function S_2^{Fe-Si} since in the low ϕ limit its values are several orders of magnitude smaller than the autocorrelation functions, which again suggests that there are no significant inter-particle spatial correlations.

By definition, the two-point cluster function C_2 measures the contributions from the point pairs in the same cluster (inclusion) in S_2; see Eq. (5). Thus, we approximate C_2 of the inclusion phase α as follows:

$$C_2^{\alpha,\beta}(r) = \phi_\alpha \left[A_\beta^\alpha \exp\left(B_\beta^\alpha \cdot r\right) + \left(1 - A_\beta^\alpha\right) \left(1 - \frac{3r}{2\bar{D}_\beta^\alpha} + \frac{r^3}{2\bar{D}_\beta^\alpha}\right) \right] \Theta\left(\bar{D}_\beta^\alpha - r\right)$$

$$(12)$$

Table 1 The values of A_β^α, B_β^α and $\bar{D}_\beta^\alpha$ for the Fe-rich inclusions along three orthogonal directions in the alloy

Fe-rich inclusions	Longitudinal	Transverse	Short transverse
A_β^α	0.801	0.850	0.902
B_β^α	−0.350	−0.504	−0.579
$\bar{D}_\beta^\alpha$	81.2	32.0	18.9

Table 2 The values of A_β^α, B_β^α and $\bar{D}_\beta^\alpha$ for the Si-rich inclusions along three orthogonal directions in the alloy

Si-rich inclusions	Longitudinal	Transverse	Short transverse
A_β^α	0.918	0.898	0.951
B_β^α	−0.480	−0.548	−0.679
$\bar{D}_\beta^\alpha$	86.8	25.9	11.0

where the parameters A_β^α, B_β^α and $\bar{D}_\beta^\alpha$ are given in Tables 1 and 2 respectively for the Fe-rich and Si-rich inclusions.

The YT stochastic reconstruction procedure is employed to generate virtual alloy microstructures from the associated S_2 [Eq. (9)] and C_2 [Eq. (12)]; see Figure 4. By quantitatively comparing the experimentally obtained structure and the reconstructions via the lineal-path functions of the inclusion phases, it can be clearly seen that the combination of S_2 and C_2 is sufficient to statistically characterize the microstructure, i.e., the complex alloy microstructure can be modeled by six directional correlation functions with closed analytical forms, which only depend on the parameters given in Tables 1 and 2. In other words, the large 3D microstructural array specifying the local state of each individual voxels can be reduced to a handful of scalar parameters that statistically model the alloy microstructure.

Polycrystalline tin solder

Tin has been widely used as solider materials in electronic packaging. Due to heterogeneous cooling and the existence of intermetallic particles, a tin solder joint usually processes a polycrystalline microstructure. The mechanical properties of the joint are largely determined by the grain size, orientation and the degree of mis-orientation between the grains. Such structural information can be obtained via electron back scattered diffraction (EBSD) experiments (see Figure 5A). However, the EBSD image technique only allows

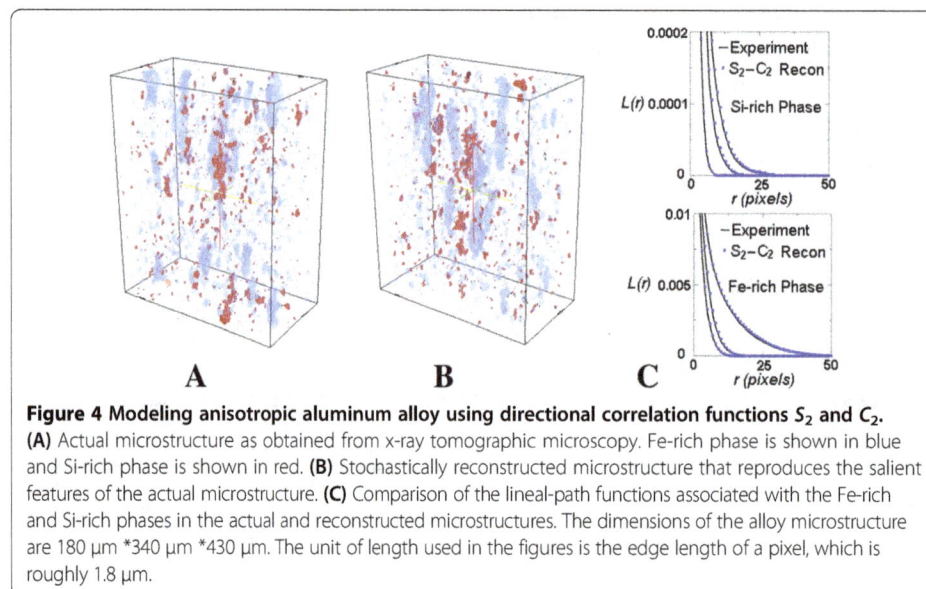

Figure 4 Modeling anisotropic aluminum alloy using directional correlation functions S_2 and C_2.
(A) Actual microstructure as obtained from x-ray tomographic microscopy. Fe-rich phase is shown in blue and Si-rich phase is shown in red. **(B)** Stochastically reconstructed microstructure that reproduces the salient features of the actual microstructure. **(C)** Comparison of the lineal-path functions associated with the Fe-rich and Si-rich phases in the actual and reconstructed microstructures. The dimensions of the alloy microstructure are 180 µm *340 µm *430 µm. The unit of length used in the figures is the edge length of a pixel, which is roughly 1.8 µm.

one to probe the surface of the material sample. Therefore, serial sectioning is required to acquire successive 2D images at different depth of the polycrystalline structure. A final reconstruction can be obtained by carefully stacking these 2D images [20,21].

Our modeling procedure provides an alternative approach to generate virtual 3D polycrystalline structure models from single or a few EBSD images of the material surface. Specifically, the colored EBSD image, in which different colors represent different grain orientations, is first converted to a gray-scale image. Then the gray-scale image is thresholded to generate a multiphase material structure such that each phase corresponds to a narrow range of gray values (i.e., a narrow distribution of orientation). The morphology of each phase includes compact regions that are associated the grains of different orientations (i.e., colors in the original EBSD image). The two-point correlation functions S_2 associated with the individual phases are computed (see Figure 5B). Under the assumption that the structure is statistically homogeneous and isotropic, the sampled S_2 from 2D images are representative of the full 3D microstructure and thus, can be used to generate 3D reconstructions. Similarly, the surface-surface correlation functions F_{SS} (r) sampled from the 2D images incorporate 3D structural information for statistically homogeneous and isotropic systems, and thus, can also be employed to render virtual 3D microstructures.

Figure 5C and D respectively shows the reconstructed 3D polycrystalline structures using S_2 alone and using the combination of S_2 and $F_{SS}(r)$. Both reconstructions statistically reproduce the key structural features including the size and shape of the grains and the

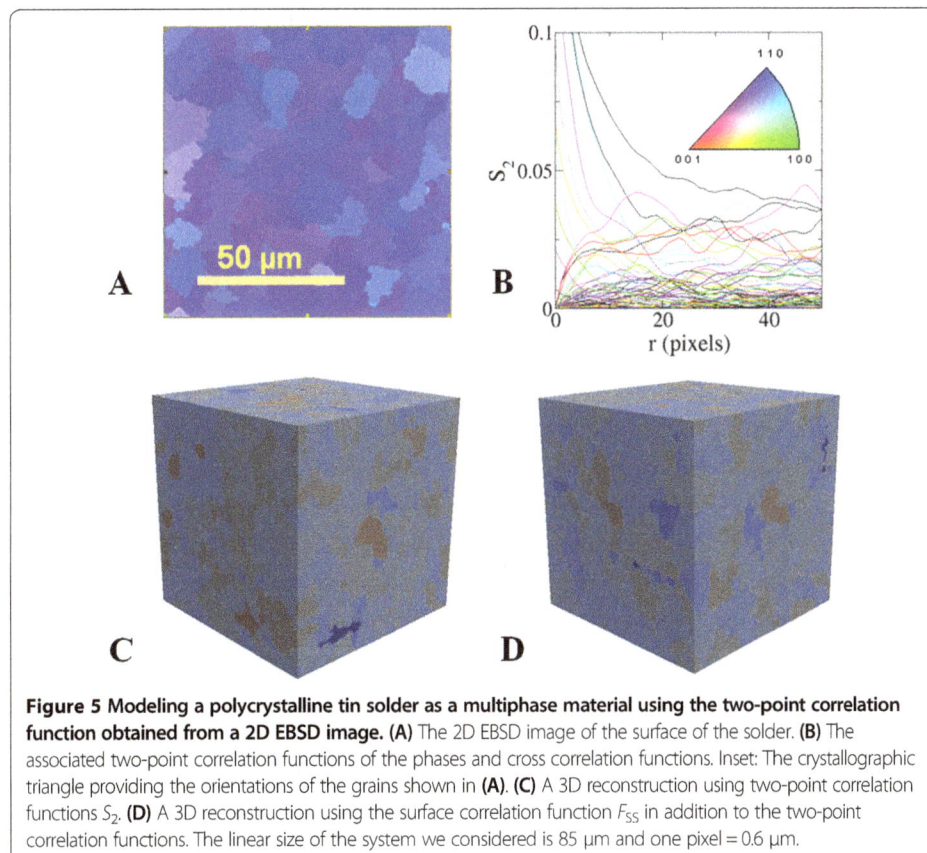

Figure 5 Modeling a polycrystalline tin solder as a multiphase material using the two-point correlation function obtained from a 2D EBSD image. (A) The 2D EBSD image of the surface of the solder. **(B)** The associated two-point correlation functions of the phases and cross correlation functions. Inset: The crystallographic triangle providing the orientations of the grains shown in **(A)**. **(C)** A 3D reconstruction using two-point correlation functions S_2. **(D)** A 3D reconstruction using the surface correlation function F_{SS} in addition to the two-point correlation functions. The linear size of the system we considered is 85 μm and one pixel = 0.6 μm.

orientation correlation between neighboring grains, as can be seen by comparing the surfaces of the 3D virtual microstructures to the target 2D image. Quantitative comparison between the reconstructions (not shown here) indicates that incorporating the surface function does not lead to additional improvement of the accuracy of the reconstruction. This suggests that S_2 alone would be sufficient to model this polycrystalline system. We emphasize that a quantitative comparison between the reconstructions and experimentally obtained 3D structure is necessary to definitely ascertain the accuracy of the reconstructions. However, the latter is currently not available.

Structural evolution in lead-tin binary alloy

Binary lead/tin alloys have been widely used as solders [48]. In particular, a eutectic alloy of 63% tin (Sn), 37% lead (Pb) has been used as interconnect due to its unique low melting point (183°C), good wettability, and excellent mechanical properties. The eutectic alloy microstructure contains a Pb-rich phase and a Sn-rich phase, which can possess both laminar and globular morphologies. The salient microstructural features such as the width and extent of the laminar phases as well as the size and spatial distribution of the globular phases can significantly affect the overall mechanical properties of the alloy. At temperatures below the eutectic melting point, the enhanced diffusion of Pb and Sn atoms can lead to significant coarsening in the alloy, which lowers the total interfacial energy [49]. A heat treatment (e.g, annealing) can then be employed to "tune" the eutectic microstructure to achieve desirable material performance. The rate of coarsening increases with temperature and it is of particular importance to quantitatively model and predict the degree of coarsening in the design of materials for high temperature applications.

Figure 6 demonstrates our modeling of the structural evolution in a binary alloy (Pb37Sn63) annealed at 175°C up to 216 h [29], which can be accurately characterized by the associated time-dependent scaled two-point correlation functions $f(r)$, i.e.,

$$f(r) = \frac{S_2(r) - \phi^2}{\phi(1-\phi)} \tag{13}$$

Using phase-field modeling, we have found that a growing length scale λ characterizing the coarsening process can be defined from the correlation length (i.e., a scalar parameter) in $f(r)$ associated with microstructures annealed for different amounts of time. 2D optical images of polished surface of the sample were taken at different annealing times (see the upper panels of Figure 6) and careful measurements of λ from these images have shown that λ satisfies the scaling relation $\lambda(t) \sim t^{1/3}$, which is consistent with the phase-field modeling results and expected for systems that are diffusion-controlled. This allows us to construct a universal functional form of $f(r)$ parameterized by $\lambda(t)$ to statistically characterize the entire spectrum of microstructures undergone coarsening, i.e.,

$$f(r) = \exp(-ar/\lambda)\cos(\pi r/\lambda + b)/\cos(b) \tag{14}$$

where $a = 3.5$ and $b = 0.6$ are parameters depending on the initial microstructure but not the annealing time. Snapshots of evolving alloy microstructure at specific annealing times reconstructed from the correlation function $f[\lambda(t)]$ are shown in the middle

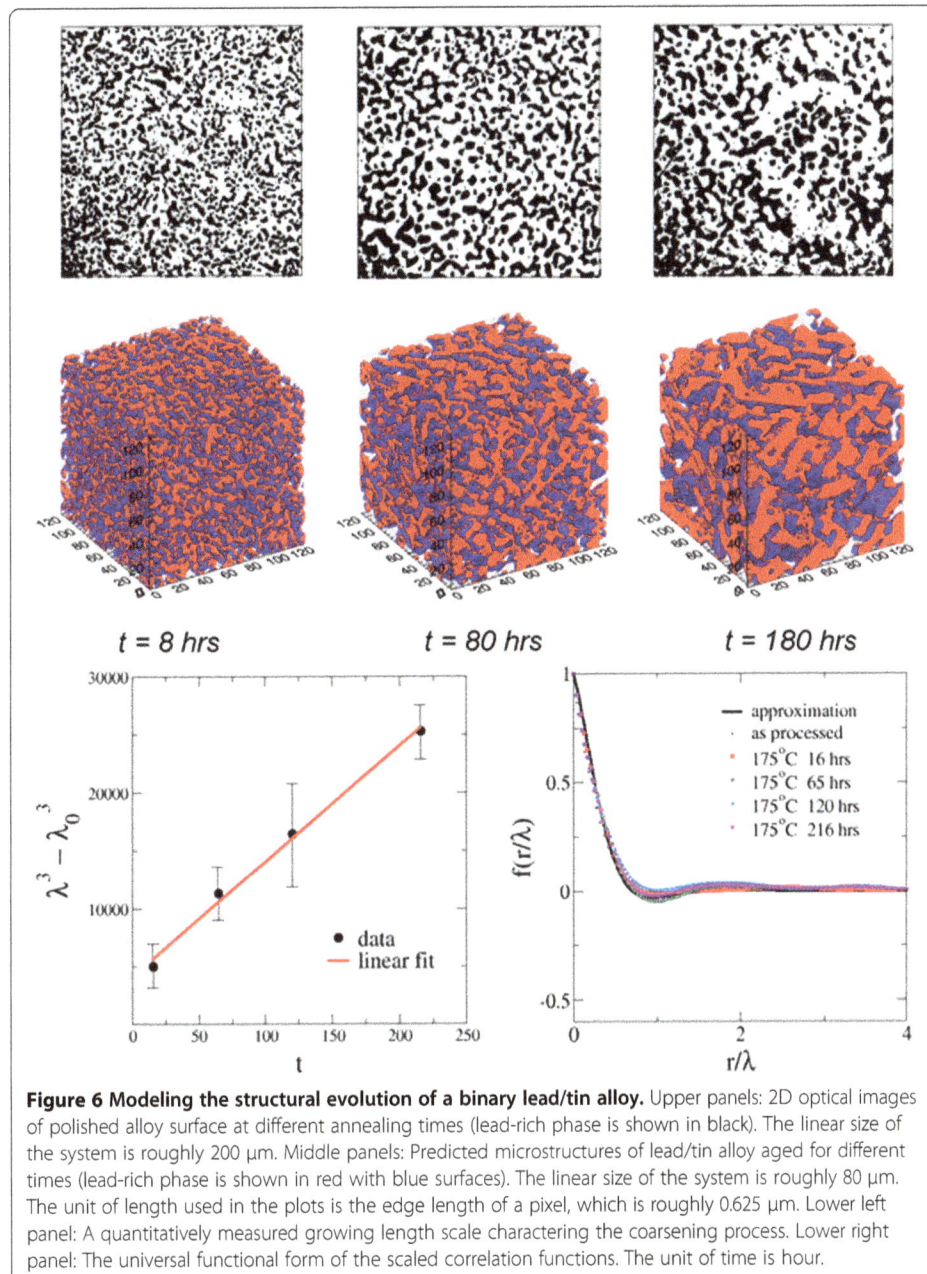

Figure 6 Modeling the structural evolution of a binary lead/tin alloy. Upper panels: 2D optical images of polished alloy surface at different annealing times (lead-rich phase is shown in black). The linear size of the system is roughly 200 μm. Middle panels: Predicted microstructures of lead/tin alloy aged for different times (lead-rich phase is shown in red with blue surfaces). The linear size of the system is roughly 80 μm. The unit of length used in the plots is the edge length of a pixel, which is roughly 0.625 μm. Lower left panel: A quantitatively measured growing length scale charactering the coarsening process. Lower right panel: The universal functional form of the scaled correlation functions. The unit of time is hour.

panels of Figure 6. The development of coarsening in the microstructure is clearly seen. The lower panels of Figure 6 show the universal function on to which the scaled f associated with experimentally obtained microstructures annealed at different times collapse. The quantitative relation between the material microstructure as represented by the correlation function $f[\lambda(t)]$ and the processing-condition parameter t lies the mathematical foundation for material design and optimization. In particular, this procedure of 4D material microstructure modeling allows one to directly and quantitatively tie processing parameters (e.g., the annealing time t in this case) to the microstructure. Subsequent analysis and simulations on the physical properties and performance can be carried out to further establish the processing-structure-performance relations.

Hard-sphere packing

Finally, we employ the procedure to examine the information content of different correlation functions associated with a hard-sphere packing, i.e., a dispersion of hard spheres in a matrix [24]. This structure has been widely used in models for particle-reinforced composites, colloids, and granular materials. The packing shown in Figure 7A is generated using the standard Metropolis Monte Carlo technique for a canonical ensemble of hard spheres in a cubical box under periodic boundary conditions. The three reconstructions shown in Figure 7B-D are generated respectively using S_2 alone, the combination of S_2 and the pore-size function F, and the combination of S_2 and the cluster function C_2. A visual comparison of the reconstruction involving C_2 reveals that it accurately yields a dispersion of well-defined spherical inclusions of the same size, in contrast to the S_2 reconstruction, which grossly overestimates clustering of the "sphere" phase. In addition, although the reconstruction incorporating F provides improvement over the rendition of the S_2-alone reconstruction, it is still inferior to the S_2–C_2 reconstruction in reproducing both the size and shape of the "sphere" phase. The accuracy of the reconstructions can also be quantitatively ascertained by comparing the lineal-path function associated with the sphere phase in the rendered microstructures, as shown in Figure 7E. This example clearly illustrates the utility of our procedure in examining the information content of various statistical descriptors for structural characterization. In the sphere-packing system, the combination of S_2 and C_2 provide a sensitive descriptor of the structure, suggesting that similar material systems (e.g., particle reinforced composites) can be accurately modeled by these quantities.

Discussion and conclusion

In this paper, we presented a framework to model the microstructure of complex heterogeneous materials via certain lower-order correlation functions associated

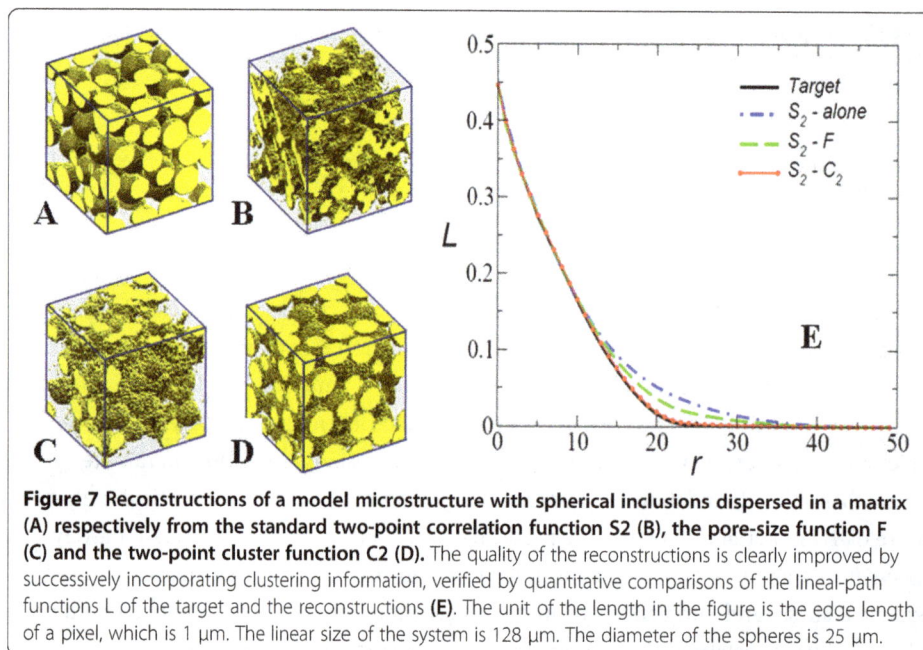

Figure 7 Reconstructions of a model microstructure with spherical inclusions dispersed in a matrix (A) respectively from the standard two-point correlation function S2 (B), the pore-size function F (C) and the two-point cluster function C2 (D). The quality of the reconstructions is clearly improved by successively incorporating clustering information, verified by quantitative comparisons of the lineal-path functions L of the target and the reconstructions **(E)**. The unit of the length in the figure is the edge length of a pixel, which is 1 μm. The linear size of the system is 128 μm. The diameter of the spheres is 25 μm.

with the material's phases. Such correlation functions quantitatively characterize various topological and geometrical features of the material microstructure and play an important role in the quantitative connection between the material microstructure and the effective physical properties as obtained via homogenization theories. Representing a microstructure with the combination of a selected set of correlation functions allows one to reduce the large data sets for a complete specification of all of the local states in a microstructure to a handful of simple scalar functions that statistically capture the salient structural features of the material. Given a specific set of correlation functions, the Yeong-Torquato reconstruction procedure can be employed to generate 3D virtual microstructure models compatible with the functions. Moreover, the Y-T procedure also allows one to examine the information content of different correlation functions by quantitatively comparing the reconstructions to the actual microstructure in order to determine the set of most sensitive morphological descriptors for such a class of materials for microstructure modeling and virtual material generation. The accuracy of the reconstruction can be quantitatively ascertained by comparing statistical descriptors (e.g., various correlation function, distribution of cluster size, continuity etc.) that are not constrained in the optimization.

To demonstrate its utility, we employed the framework to model a variety of material systems, including an aluminum alloy with anisotropic secondary phases, a polycrystalline tin solder, the structural evolution in binary lead-tin alloy when aged, and a model microstructure with hard spheres dispersed in a matrix. We found that for certain materials, such as the lead-tin alloy, the standard two-point correlation function S_2 alone is sufficient to provide a statistically accurate characterization of the associated microstructure. On the other hand, a realistic reconstruction of the hard-sphere packing model clearly requires additional topological connectedness information as contained in the cluster function C_2. In general, we expect that successively incorporating additional morphological information by increasing the number of correlation functions as input for the Y-T procedure would lead to more and more accurate reconstructions. In practice, a set of carefully chosen correlation functions sensitive to the key structure features would be sufficient for characterizing and modeling a material of interest. We also note that for certain microstructures such as percolated thin filaments, topological transformations (e.g., dilation and erosion) could be used to further improve the accuracy of the reconstruction [50].

Our framework enables one to model structural evolution over time via time-dependent correlation functions. Physics-based models (e.g., the phase-field modeling in the case of the lead-tin alloy example shown here) can be employed to suggest growing length scale or other key structural parameters that are directly manifested in the correlation function modeling the material structure. A quantitative relation between the processing condition and the structural evolution is established by incorporating the key processing parameters such as annealing temperature and annealing time into the associated correlation functions. Such relations are crucial to the integrated design of optimal multifunctional materials. Besides examining the information content of correlation functions, the stochastic reconstruction procedure can be adapted to investigate other types of structure data, such as limited-angle tomography projections. Understanding the information content of tomography data could lead to significant reduction of the number of projections acquired to characterize a single microstructure and thus, improve the efficiency for 4D characterization using tomography.

Availaiblity of supporting data

The C++ program for the stochastic reconstruction described in the paper is available upon request.

Competing interests
The authors declare no competing interests.

Authors' contributions
YJ and NC contributed equally to this article. Both authors read and approved the final manuscript.

Acknowledgements
This work was supported by the Division of Materials Research at National Science Foundation under award No. DMR-1305119. The authors acknowledge A. Kirubanandham for the EBSD images of pure Sn. Y. J. also thanks Arizona State University for the generous start-up package.

References

1. Christensen RM (1979) Mechanics of Composite Materials. Wiley, New York, 1979
2. Nemat-Nasser SM, Hori M (1999) Micromechanics: Overall Properties of Heterogeneous Solids. Elsevier Science Publishers, Amsterdam
3. Torquato S (2002) Random Heterogeneous Materials: Microstructure and Macroscopic Properties. Springer, New York
4. Sahimi M (2003) Heterogeneous Materials I: Linear Transport and Optical Properties, and II: Nonlinear and Breakdown Properties and Atomistic Modeling. Springer, New York
5. Millar DIA (2012) Energetic Materials at Extreme Conditions. Springer Ph.D. Thesis, New York
6. Haymes RC (1971) Introduction to Space Science. John Wiley and Sons Inc, New York, NY, 1971
7. Thornton K, Poulsen HF (2008) Three-dimensional materials science: an intersection of three-dimensional reconstructions and simulations. MRS Bull 33:587
8. Brandon D, Kaplan WD (1999) Microstructural Characterization of Materials. John Wiley & Sons, New York
9. Baruchel J, Bleuet P, Bravin A, Coan P, Lima E, Madsen A, Ludwig W, Pernot P, Susini J (2008) Advances in synchrotron hard x-ray based imaging. C R Physique 9:624
10. Kinney JH, Nichols MC (1992) X-ray tomographic microscopy (XTM) using synchrotron radiation. Annu Rev Mater Sci 22:121
11. Kak A, Slaney M (1988) Principles of Computerized Tomographic Imaging. IEEE Press, New York
12. Babout L, Maire E, Buffière JY, Fougères R (2001) Characterisation by X-ray computed tomography of decohesion, porosity growth and coalescence in model metal matrix composites. Acta Mater 49:2055
13. Borbély A, Csikor FF, Zabler S, Cloetens P, Biermann H (2004) Three-dimensional characterization of the microstructure of a metal-matrix composite by holotomography. Mater Sci Engng A 367:40
14. Kenesei P, Biermann H, Borbély A (2005) Structure–property relationship in particle reinforced metal-matrix composites based on holotomography. Scr Mater 53:787
15. Weck A, Wilkinson DS, Maire E, Toda H (2008) Visualization by x-ray tomography of void growth and coalescence leading to fracture in model materials. Acta Mater 56:2919
16. Toda H, Yamamoto S, Kobayashi M, Uesugi K, Zhang H (2008) Direct measurement procedure for three-dimensional local crack driving force using synchrotron X-ray microtomography. Acta Mater 56:6027
17. Williams JJ, Flom Z, Amell AA, Chawla N, Xiao X, De Carlo F (2010) Damage evolution in SiC particle reinforced Al alloy matrix composites by X-ray synchrotron tomography. Acta Mater 58:6194
18. Williams JJ, Yazzie KE, Phillips NC, Chawla N, Xiao X, De Carlo F, Iyyer N, Kittur M (2011) On the correlation between fatigue striation spacing and crack growth rate: a three-dimensional (3-D) X-ray synchrotron tomography study. Metall Mater Trans 42:3845
19. Williams JJ, Yazzie KE, Phillips NC, Chawla N, Xiao X, De Carlo F Understanding fatigue crack growth in aluminum alloys by in situ x-ray synchrotron tomography. Int J Fatigue, in press
20. Groeber M, Ghosh S, Uchic MD, Dimiduk DM (2008) A framework for automated analysis and simulation of 3D polycrystalline microstructures. I: statistical characterization. Acta Mater 56:1257
21. Groeber M, Ghosh S, Uchic MD, Dimiduk D (2008) A framework for automated analysis and simulation of 3D polycrystalline microstructures. II: synthetic structure generation. Acta Mater 56:1274
22. Jiao Y, Stillinger FH, Torquato S (2007) Modeling heterogeneous materials via two-point correlation functions: basic principles. Phys Rev E 76:031110
23. Jiao Y, Stillinger FH, Torquato S (2008) Modeling heterogeneous materials via two-point correlation functions: II. Algorithmic details and applications. Phys Rev E 77:031135
24. Jiao Y, Stillinger FH, Torquato S (2009) A superior descriptor of random textures and its predictive capacity. Proc Natl Acad Sci 106:17634
25. Yeong CLY, Torquato S (1998) Reconstructing random media. Phys Rev E 57:495

26. Yeong CLY, Torquato S (1998) Reconstructing random media: II. Three-dimensional reconstruction from two-dimensional cuts. Phys Rev E 58:224

27. Liu Y, Greene MS, Chen W, Dikin DA, Liu WK (2013) Computational microstructure characterization and reconstruction for stochastic multiscale material design. Computer-Aided Design 45:65

28. Mikdam A, Belouettar R, Fiorelli D, Hu H, Makradi A (2013) A tool for design of heterogeneous materials with desired physical properties using statistical continuum theory mater. Sci Eng A 564:493

29. Jiao Y, Pallia E, Chawla N (2013) Modeling and predicting microstructure evolution in lead/tin alloy via correlation functions and stochastic material reconstruction. Acta Mater 61:3370

30. Roberts AP (1997) Statistical reconstruction of three-dimensional porous media from two- dimensional images Phys. Rev E 56:3203–12

31. Niezgoda SR, Kanjaria RK, Kalidindi SR (2013) Novel microstructure quantification framework for databasing, visualization, and analysis of microstructure data. Inter Mater Manu Innov 2:3

32. Niezgoda SR, Fullwood DT, Kalidindi SR (2008) Delineation of the space of 2-point correlations in a composite material system. Acta Mater 56:5285–92

33. Lu B, Torquato S (1992) Lineal path function for random heterogeneous materials. Phys Rev A 45:922

34. Lu B, Torquato S (1992) Lineal path function for random heterogeneous materials II. Effect of polydispersivity. Phys Rev A 45:7292

35. Torquato S, Beasley JD, Chiew YC (1988) Two-point cluster function for continuum percolation. J Chem Phys 88:6540

36. Cinlar E, Torquato S (1995) Exact determination of the two-point cluster function for one-dimensional continuum percolation. J Stat Phys 78:827

37. Torquato S (1986) Interfacial surface statistics arising in diffusion and flow problems in porous media. J Chem Phys 85:4622

38. Torquato S, Avellaneda M (1991) Diffusion and reaction in heterogeneous media: pore-size distribution, relaxation rimes, and mean survival time. J Chem Phys 95:6477

39. Prager S (1963) Interphase transfer in stationary two-phase media. Chem Eng Sci 18:228

40. Fullwood DT, Niezgoda SR, Kalidindi SR (2008) Microstructure reconstructions from 2-point statistics using phase-recovery algorithms. Acta Mater 56:942–48

41. Hajizadeh A, Safekordi A, Farhadpour FA (2011) A multiple-point statistics algorithm for 3D pore space reconstruction from 2D images. Advances in Water Resources 34:1256–1267

42. Tahmasebi P, Sahimi M (2013) Cross-correlation function for accurate reconstruction of heterogeneous media Phys. Rev Lett 110:078002

43. Sheehan N, Torquato S (2001) Generating microstructures with specified correlation functions. J Appl Phys 89:53

44. Rozman MG, Utz M (2002) Uniqueness of reconstruction of multiphase morphologies from two-point correlation functions. Phys Rev Lett 89:135501

45. Kirkpatrick S, Gelatt CD, Vecchi MP (1983) Optimization by simulated annealing. Science 220:671–80

46. Singh SS, Williams JJ, Jiao Y, Chawla N (2012) Modeling anisotropic multiphase heterogeneous materials via directional correlation functions: simulations and experimental verification metall. Mater Trans 4A:4470–4474

47. Debye P, Anderson HR, Brumberger H (1957) Scattering by an inhomogeneous solid. II. The correlation function and its applications. J Appl Phys 28:679–83

48. Manko HH (2001) Solders and Soldering: Materials, Design, Production, and Analysis for Reliable Bonding. McGraw-Hill, New York

49. Porter DA, Easterling KE (2004) Phase Transformations in Metals and Alloys. Taylor & Francis, London

50. Guo EY, Chawla N, Jing T, Torquato S, Jiao Y (2014) Accurate modeling and reconstruction of three-dimensional percolating filamentary microstructures from two-dimensional micrographs via dilation-erosion method. Mater Chara 89:33–42

2

Integrated computational materials engineering from a gas turbine engine perspective

Ann Bolcavage[1], Paul D Brown[2], Robert Cedoz[1], Nate Cooper[1], Chris Deaton[1], Daniel R Hartman[1], Akin Keskin[2], Kong Ma[1], John F Matlik[1*], Girish Modgil[1] and Jeffrey D Stillinger[1]

* Correspondence:
John.F.Matlik@rolls-royce.com
[1]Rolls-Royce Corporation,
Indianapolis, IN 46225, USA
Full list of author information is
available at the end of the article

Abstract

In 2008, the National Research Council published a landmark report on Integrated Computational Materials Engineering (ICME) and defined it as 'an emerging discipline that aims to integrate computational materials science tools into a holistic system that can accelerate materials development, transform the engineering design optimization process, & unify design and manufacturing'. ICME is becoming a critical enabler for reducing the design/make cycle time and getting complex systems into production more quickly. There are several reasons why this is the case. Firstly, ICME allows materials experts to develop new material systems and methods of manufacture much more quickly. Advanced new materials and their associated manufacturing processes can be tailored to deliver products that meet design requirements quickly and more effectively in terms of cost and performance. Secondly, ICME enables design processes to quantify cause and affect relationships between manufacturing methods and variability, material properties, product geometry, and design requirement margins. In the design phase, material selection itself can impose consideration of material-specific failure modes that are naturally correlated to important attributes such as strength, weight, and geometry. ICME enables designers to quickly understand the complex and probabilistic interactions between the material, manufacturing processes, manufacturing variability, and design. Thirdly, it has been shown that successful account of variability of the manufacturing processes in life calculations leads to improved accuracy in declared low cycle fatigue crack initiation and damage tolerance lives on life limited gas turbine engine components. Furthermore, ICME enables engineers to rapidly explore more effective design and manufacturing solutions for delivering superior products at lower cost, faster but not without challenges. To highlight challenges and progress toward realization of this transformational technology, a survey of recent examples of materials and manufacturing process simulations along with the overarching approach and requirements within ICME to link these simulation capabilities to design and manufacturing methods will be reviewed from a gas turbine engine perspective.

Keywords: Integrated computational materials engineering; Materials and process modeling; Integrated design and make; Design systems; Gas turbine engine

Review

Impetus - why we care

System benefits are lost when communication and technical data does not pass across functional lines. System failures occur when 'islands' or 'silos' are allowed to persist in product development. Fundamentally, the maturing discipline of Integrated Computational Materials Engineering (ICME) aims to integrate more than 50 years of computational materials and process modeling research into engineering systems to accelerate the design/make process through use of physics-based models and virtual linkage of the manufacturing processes with material structure and design performance. Though historical focus has been on development of digital/virtual toolsets, the success and practical implementation of ICME into product development systems requires and forces the cross-functional information flow and cultural change that is required to realize significant manufacturing benefit and improve competitiveness. Digitally integrating design and manufacture up front for iteration in analysis is much less costly than redesigning and replacement of products in service.

Successful integration of ICME into product development systems promises order of magnitude reductions in product development time and cost through use of physics-based models which replace costly experimental iterations for optimizing manufacturing processes and component performance, as illustrated in Figure 1. The magnitude of benefit promised through systems level solutions, ICME, and the necessity for a step change in manufacturing competitiveness has primed industry for the next manufacturing revolution. In addition to the reduction in product development cost and timeline for engine development, life cycle cost benefits can be realized after entry into service.

Figure 2 illustrates the benefits of an improved approach to product development by introducing ICME earlier in the *design stage* after incorporating all relevant knowledge throughout the product life cycle. It can shorten the time to market and product capability responsiveness and deliver savings through reduction of recurring production and part costs. Improved accuracy in declared lives for life-limited components derived by the application of physics-based modeling of manufacturing processes can provide a significant reduction in cost of operation and maintenance of gas turbine engines. In addition, the application of ICME with Prognostic Health Monitoring (PHM) can also deliver significant business benefit by better informing maintenance decisions and overall fleet risk evaluations. Integration and application of computational materials

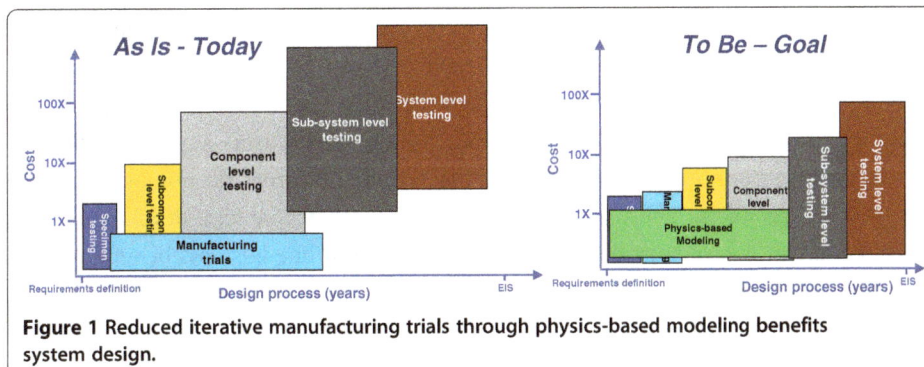

Figure 1 Reduced iterative manufacturing trials through physics-based modeling benefits system design.

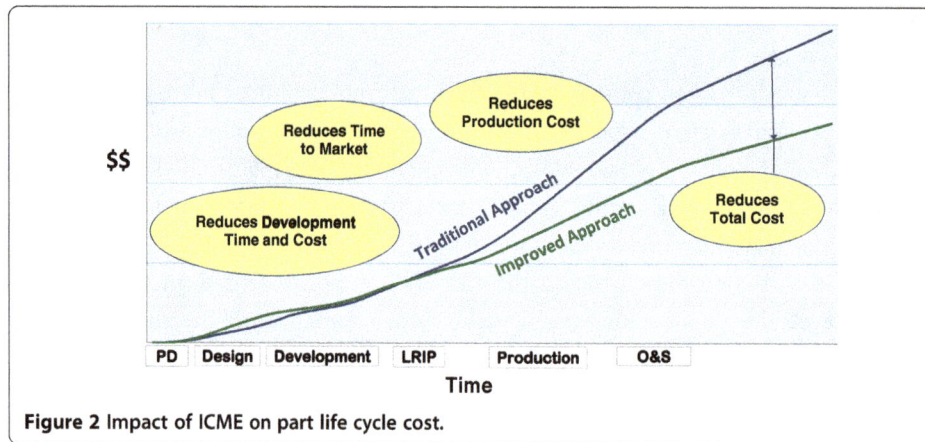

Figure 2 Impact of ICME on part life cycle cost.

engineering capability at Rolls-Royce has delivered significant business benefit in a number of ways including, but not limited to the following:

- Distortion prediction of forged parts resulting in less scrap, fewer concessions (99.98% Right First Time), decreasing forging machining costs (50% cost reduction), and increasing stock-turn.
- Reduction in casting scrap (> > US$5 M p.a.).
- Increasing tensile strength (approximately 5% on forged/formed parts).
- Significant reduction (by 90%) of forming trials (powder HIP process) due to predictive part geometric accuracy.
- Optimization of parts in furnaces to increase utilization; increasing stock-turn and lowering costs.
- Shortening development time and time to market for new product introduction and redesign efforts.

The magnitude and opportunity of the benefit afforded by successful integration of ICME into product development systems is substantial. However, the implementation of ICME into product development systems has several significant challenges that must be addressed if bottom line business benefit is to be fully realized.

Scoping the ICME landscape

The integration of computational materials and process modeling into product development systems promises significant benefit for cost reduction, performance improvement, and lead time reduction across many material systems, commodity streams, and products, as well as across the product life cycle itself. As a result, a key challenge exists in appropriately focusing ICME efforts to match the greatest business benefit rather than trying to solve too broad a problem. Manufacturing companies do not sell technology; they sell products. Thus, all prioritization of ICME technology development and implementation efforts must have line-of-sight to bottom line business benefit and delivering superior product to market faster and at lower cost. ICME efforts can fail before they start simply because they have not been scoped for success and the focus falls solely on developing technology and not delivering to customer requirements and business benefit. Initially, internal design and manufacturing functions within the business can be engaged to articulate how ICME methods and toolsets can better deliver to their greatest product improvement needs. This facilitates prioritization of efforts to specific prioritized product benefits, delivered through specific components or subsystems, and identifies specific design/make stakeholders that will become future business champions if the project is successful. For example, Figure 3 illustrates some general benefits/needs internal sub-system teams might articulate for a gas turbine [1].

In addition to continual customer engagement, it is important to consider several other scoping factors for ICME effort prioritization. The selected effort must be planned to deliver a phased benefit with a series of quick wins early on to maintain business support. This may mean that the selected project may be the one with the greatest business impact, or the project that delivers the largest benefit for an acceptable level of risk. Furthermore, overlap in benefits or needs articulated from customers may highlight efforts that would have a broader impact in areas across the business. Wherever possible, planning to develop once and reuse multiple times will pay big dividends when scaling up the benefit and may highlight a lower risk project that can be selected as a pilot while providing the foundation for future benefit on higher value components and systems. For example, engagement with design, component structural analysis and lifing, and materials and manufacturing functions internally at Rolls-Royce led to grouping ICME's capability into commodity buckets with similar manufacturing process applicability as highlighted in the tabs of Figures 4 and 5 [2].

These process modeling maps provide a graphical illustration implying the sequence of ICME models that virtually link design requirements, the manufacturing processes that make up a product's 'digital pedigree,' and the predicted material structures and properties that can predict the component manufacturing yield and design performance. This gives context for key stakeholders to quickly contextualize the larger ICME landscape for a given commodity group and highlights a digital thread for each component that links back to the design requirement articulated from customer engagement.

It is important to note that the integration of ICME into product development systems is not just digital but physical, heuristic, and relational as well. As a gas turbine engine OEM, Rolls-Royce delivers world-class power system and service solutions to our customers... but not on our own. This requires a trusted ICME supply chain and infrastructure of incentivized technology partners that can collaborate to deliver what is best for Rolls-Royce, as an original equipment manufacturer (OEM), and for

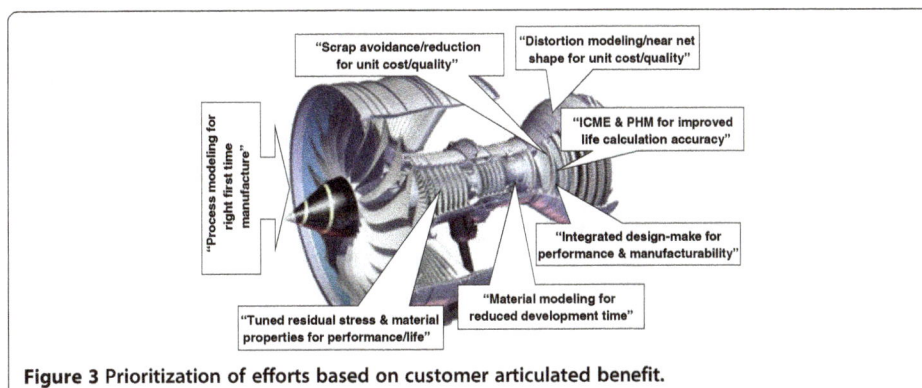

Figure 3 Prioritization of efforts based on customer articulated benefit.

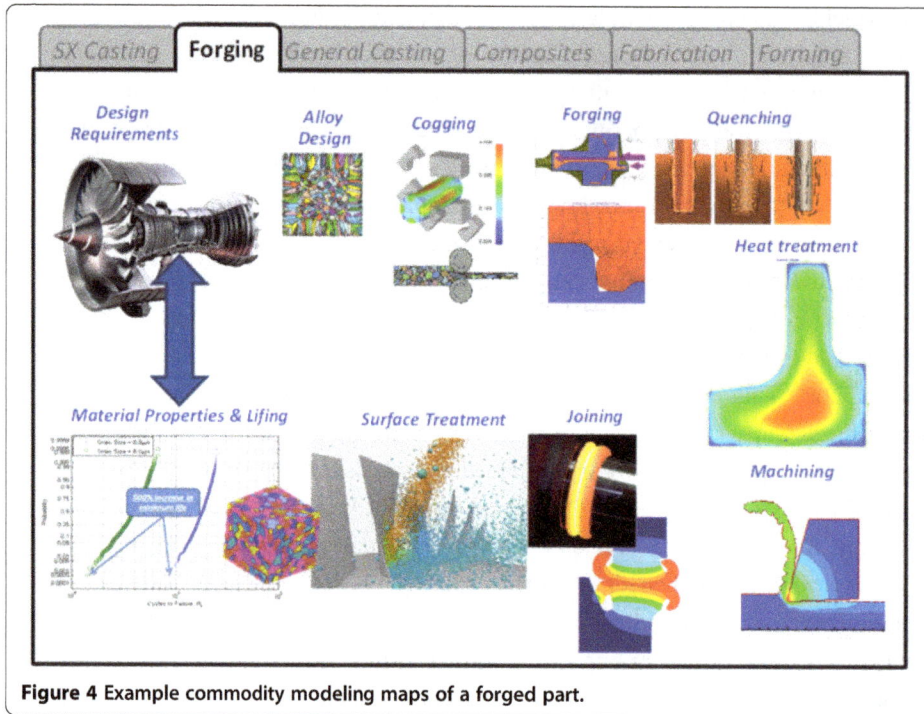

Figure 4 Example commodity modeling maps of a forged part.

the product development team as a whole. The interdisciplinary nature of systems engineering underlines the criticality of selecting an appropriate cross-functional team (i.e., design, materials, manufacturing, supply chain, and export control/IP) for program definition up front to ensure that programs are planned to deliver the right information to the right product development team members at the right time.

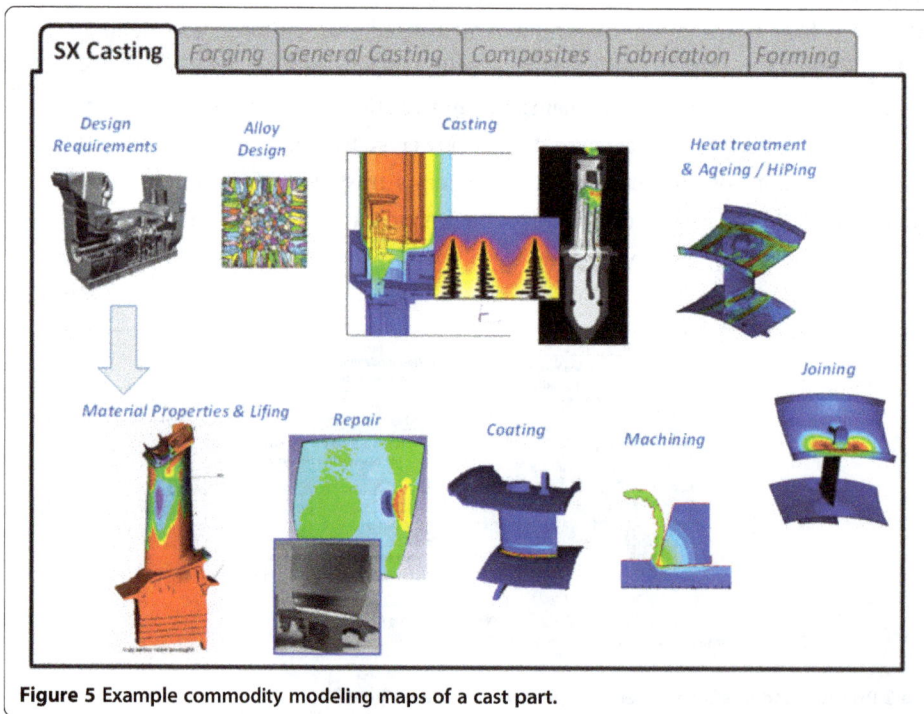

Figure 5 Example commodity modeling maps of a cast part.

Though ICME concepts are not new to the aerospace industry [3] and academic work to develop isolated computational material models has been going strong for several decades [4], the competency of the existing supply chain in integration and use of ICME methods and capability is still relatively immature. To facilitate more rapid adoption of this technology across the industry, government and industry must fund, sponsor, and partner in activities aligned to greatest business benefits. To this end, Rolls-Royce continues to invest in developing and deploying new technology through strategic partnerships with a global University Technology Center network, Advanced Manufacturing Research Centers, and public-private partnerships like the Metals Affordability Initiative [5]. An illustration of the strategic partners and development and deployment activities for a healthy ICME supply chain are highlighted in Figure 6.

Systems integration

To further scope the ICME opportunity and challenge, it is important to understand that ICME is not and should not be restricted to only design. For example, Rolls-Royce has a vision to embed ICME across the entire product development, introduction, and life cycle management process, including the following:

- Preliminary design (more analysis at sub-assembly and engine level to eliminate risk early in the design process).
- Detailed design (use of analysis-based optimization and robust design and manufacture to enable rapid definition at component and sub-system level).
- System design using high-fidelity virtual engine (more sophisticated, multiphysics analysis to accurately predict engine behavior linking manufacturing variability to component-specific performance).
- Virtual manufacture (optimization of those aspects of manufacturing that have no impact on the product design but may affect, for example, the cost).
- Virtual testing (analysis-based test strategy, planning, and correlation to reduce the need for repeat testing).

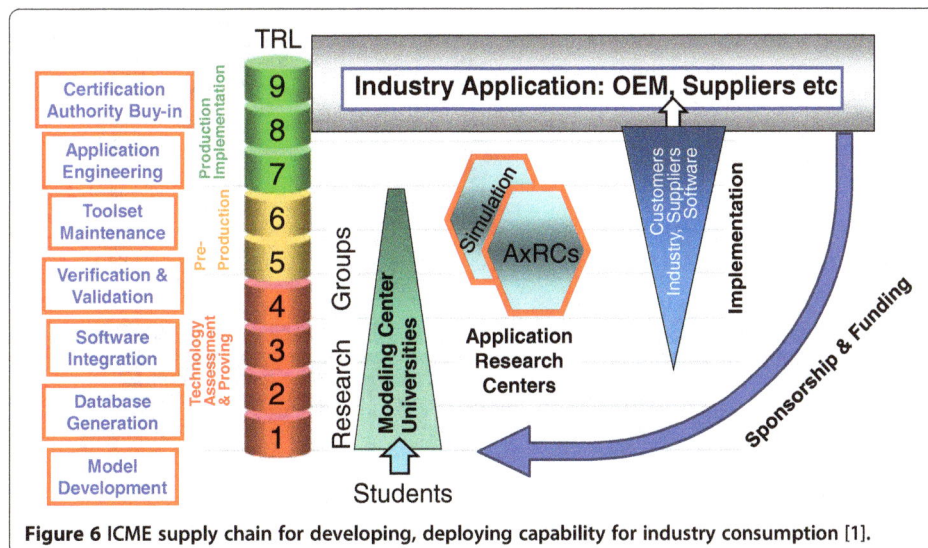

Figure 6 ICME supply chain for developing, deploying capability for industry consumption [1].

- Virtual product validation/certification (rapid certification based on validated analysis, simulation and modeling).
- Product life cycle analysis (fixed design, probabilistic analysis, and updating of models used during development for improved service and aftermarket decisions involving business risk, such as maintenance and product improvement costs).
- Process capability analysis (data and knowledge capture for reuse in continuously validated and improved methods, constraints, and rules used above).

These describe the elements of larger virtual product development system for faster delivery of superior products to market at lower cost. Figure 7 illustrates a 'six-stage' process defining the life cycle of a product and how each of these elements roughly maps onto it.

As illustrated by the dashed lines in Figure 7, both legacy and new product development efforts generate significant data and knowledge that can be leveraged in unique and innovative ways at different points in the life cycle to deliver benefit. However, with the addition of more complex analysis, the accompanying large volume of digital data that will be generated out of such complex systems and the pragmatic reality that uncertainty will pervade data and model inputs, the ability to manage risk and effectively utilize these new capabilities to inform engineering decisions requires an appropriate framework and mindset.

Facilitating engineering decision making: informing decisions under conditions of complexity and uncertainty

Managing the development of complex new systems requires many decisions using combinations of analytical engineering models, experimental data, and expert knowledge. This process is inherently expensive when risks are high and is made even more expensive when risks are not effectively managed throughout the process. Though new systems may ultimately be successful in delivering product to market, the path to success is often far from ideal in terms of the use of models, data, and knowledge to manage risk. Models, data, and knowledge are all valuable resources in their own way but are often pitted

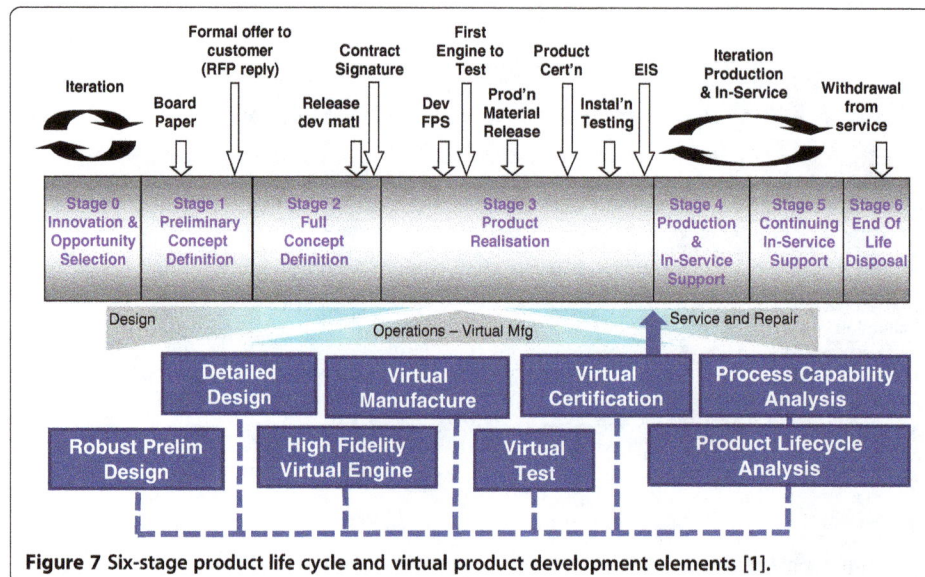

Figure 7 Six-stage product life cycle and virtual product development elements [1].

against one another inappropriately at key decision points or worse, trusted implicitly or ignored altogether. Decisions must be made, for example, about what to analyze, what to test, what conditions to test, whether a new technology should be inserted, and the level of model fidelity needed.

Significant reduction in the development time and cost depends on quantifying and responding appropriately to risk as early during the process as possible. An error in material selection or manufacturing process definition caught early during design is not nearly as costly as the same error caught later. There are, of course, several reasons such an error might not be caught early. First, the breadth in ramifications may simply not be taken into account. For example, the material may be selected based on a requirement to operate at higher temperatures but other requirements such as corrosion resistance, manufacturability, and inspectability may not be adequately considered or known. Secondly, an understanding of the typical behavior of the material or manufacturing process may be based on relatively small samples or idealized conditions. Later when the true variability becomes apparent, it becomes much more difficult to manage. In both examples, decisions are made under conditions that are unnecessary given many of the tools available today. Physics-based models such as those envisioned under ICME are being used to quantify not only typical behavior but also variability. In addition, today's access to affordable computational resources is making uncertainty quantification (UQ) for large complex systems possible when addressing the breadth issue.

However, quantification and management of risks associated with complex new systems mean making UQ and decision-making tools and training available to the right people throughout the process. Uncertainty and therefore risk tends to be highest at the onset of development and will be continually managed over time. What is not as well appreciated is the impact of sub-optimal paths on cost. The path used to blend and continually update engineering models, experimental data, and expert knowledge into information useful for decision making is critical to reducing development time and cost.

Finally, the same arguments made for reducing development costs may be advanced for reducing life cycle costs. Despite the enormous investments made to reduce risk and uncertainty prior to production release, a great deal is still learned over the life cycle of the product. New data is gathered on product performance, operational usage, changing economic conditions, for example, which continually present challenges to the decision-making process. Win-win situations can be made for OEMs, suppliers, and customers alike simply through the effective use of updating methods (e.g., Bayesian or otherwise) and decision-making tools.

ICME in context

Successful integration of ICME into next-generation product development systems will involve distributed (i.e., cross-supply chain) problem-solving solutions the ICME supply chain can *understand*, *trust*, and *use* efficiently and effectively to deliver a superior product to market faster and at lower cost.

Understand it

ICME system elements must be developed and delivered in a context and framework that product development teams can understand. Figure 8 highlights some of the key system elements involved in a product development system.

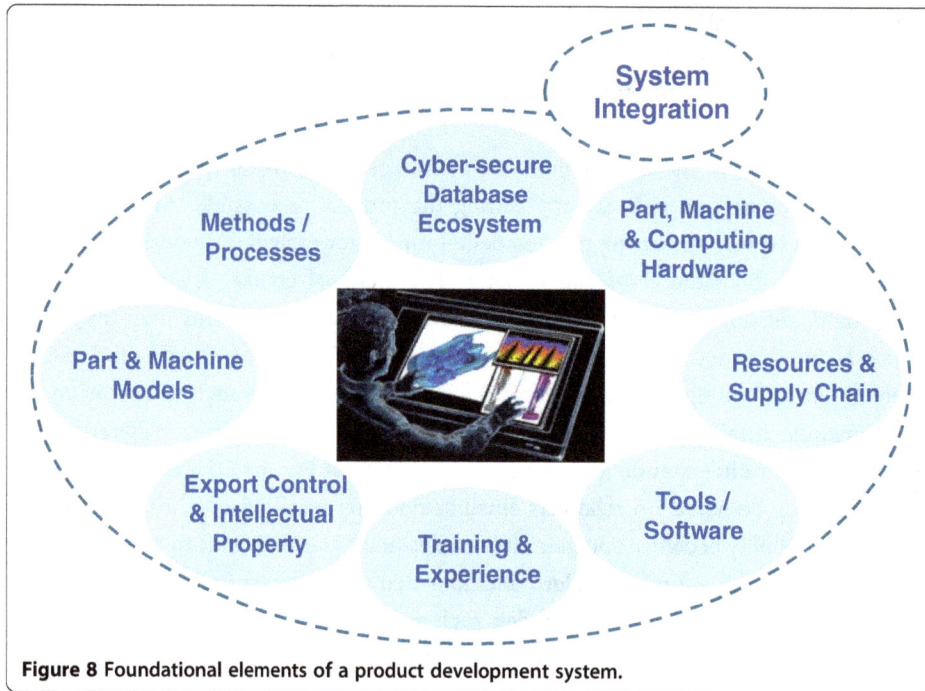

Figure 8 Foundational elements of a product development system.

During the design process, a cross-functional integrated product development team will exercise these system elements to develop and deliver a product solution. To contextualize how capability is delivered into a digitally integrated design and make system, consider the high-level engineering workflow at the sub-system level of a forged component highlighted in Figure 9. Figure 9 emphasizes the cross-functional and cross-supply chain interfaces that must be managed efficiently and effectively to facilitate truly concurrent engineering earlier in the product development process.

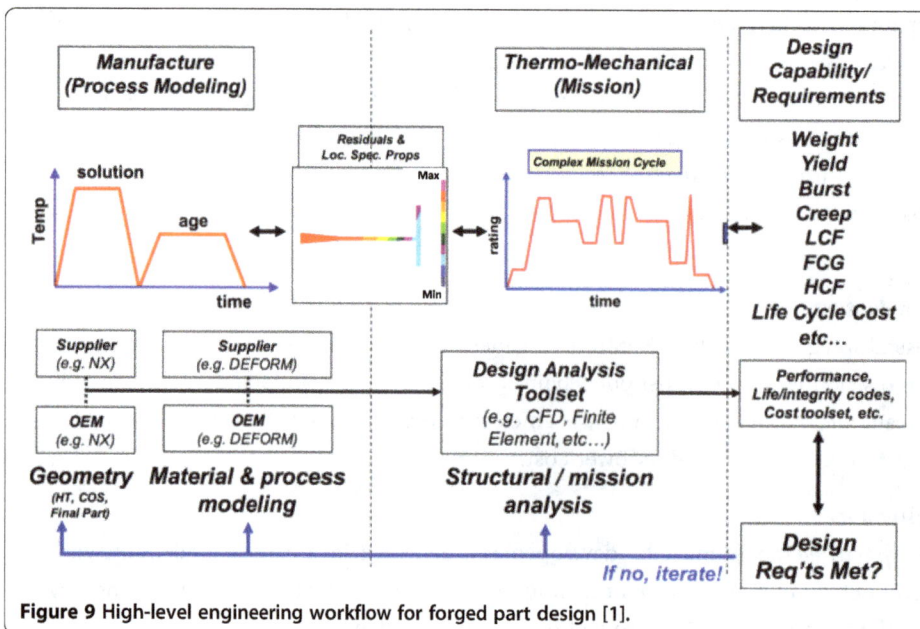

Figure 9 High-level engineering workflow for forged part design [1].

Computational material models and process models deliver predicted component performance in terms of residual stress and location-specific material properties that now enable designers to better understand the impact of manufacturing processes and manufacturing capability on the resulting design space. Digital integration and automation of the engineering workflow enables rapid exploration of feasible design space (based on known material and manufacturing capability) while simultaneously identifying robust methods of manufacture to deliver 'Right First Time' solutions that meet the requirements while minimizing or eliminating expensive physical trials. The integration of new ICME capability into workflows for components will require additional work by the engineering function in an environment that already has extreme time pressure to deliver results. Thus, ICME development efforts must address a number of design requirements and needs, or time pressure and frustration will prevent use and realization of benefits associated with the additional capability. To this end, ICME models, methods, and software toolsets must be as follows:

- *Accurate enough* Inaccuracy can result in unanticipated life cycle costs, but too much attention to accuracy results in overly high development costs. Both ultimately represent risk that needs to be quantified and managed. This is one area in which uncertainty quantification, ICME, and decision analysis tools become particularly useful. Early in the product life cycle when understanding system, sub-system, and component level requirements is especially challenging, such tools enable key decision makers to quantify what is 'accurate enough' to move forward in the development process. Accuracy must be considered both qualitatively and quantitatively. The qualitative component of accuracy (the pattern of behavior) must be consistent with physical reality to appropriately capture trending. Quantitative accuracy depends on several things including, but not limited to, the modeling algorithm, input data accuracy, material/process variations, and measurement accuracy for validation efforts. In some cases, in-process sensing and control can be used to compensate/mitigate quantitative inaccuracy. Model development efforts should be limited to the accuracy needed for the specific product or application of interest. A clear understanding of the requirements up front will define what accuracy is 'fit for purpose' and scope development efforts to deliver an appropriately refined engineering solution.
- *Efficient and usable* Engineers are fighting a battle against timescales for program delivery, and the inclusion of new capability is asking them to do more with no additional calendar time. Therefore, new capability must be delivered for seamless integration into existing design systems with as little additional computational and/or calendar time as possible.
- *Relevant* Though most material models must operate at smaller length scales (micro- or nanoscale) to capture appropriate mechanisms, delivered solutions must be compatible with continuum scale which is where most engineering analysis occurs.
- *Available* Export control and intellectual property implications with sharing models and data must be reviewed and planned for up front. This ensures that the developed solution has all required business agreements and export licenses in place for the cross-supply chain product development team while fully complying with export regulations and legal arrangements. Furthermore, a cyber-secure

information technology collaboration platform/solution must be available to streamline access to models and data while ensuring export and intellectual property right compliance.

- *Validated* Technical and business benefit validation must be considered. Manufacturing variability and uncertainty must be addressed to build confidence and trust in predicted results; however, validation of the ICME capability to deliver bottom line business benefit is what will transition from technology push to business pull.
- *Maintainable* Developed solutions must be maintained, version-controlled and updated as improvements are made or system upgrades occur. A competent and robust ICME supply chain is required to support maintenance of developed solutions or they will cease to be useful.
- *Interoperable* There are many different product development solutions and software used across the supply chain currently. This poses a challenge for streamlining digital communication between each organization. To this end, ICME efforts must deliver generic, transferrable, and modular solutions that are cross-supply chain compatible and not OEM or supplier specific.

Trust it - verification and validation

Manufacturing variability and uncertainty throughout product development is real and inevitable. Not everything is controlled and/or known which leads to substantial risk if not acknowledged and managed accordingly. In addition to verifying developed solutions to give expected outputs, validation that predicted results line up with reality requires an account of variability and uncertainty inherent to the manufacturing process.

In line with best practice robust design and make principles, it is important to identify, prioritize, and quantify sources of variation and uncertainty that impact key process variables (KPVs) in manufacturing, material structure, and ultimately product characteristics and performance in design. Conventional robust design toolsets (e.g., fishbone diagrams, p-diagrams) can be used to help engineering teams identify sources of variability and uncertainty. Expert opinion, available prior knowledge/data, and statistical models and tools can then be used to prioritize key drivers for performance and cost (i.e., Pareto chart, QFD). As is often the case when uncertainty is high, expert opinion needs to be combined carefully with analysis models and available data. This can be done working within a Bayesian statistical framework which then provides a general foundation for incorporating new data, driving toward reduced uncertainty along with verification and validation. The use of ICME toolsets and strategic testing can be leveraged to formally quantify, understand, and reduce variability and uncertainty to improve results to desired customer goal with an acceptable level of confidence. Figure 10 graphically illustrates the general process steps of manufacturing variability and uncertainty quantification and reduction.

Comprehensive implementation of ICME, especially at the engine system level, will require systematic, rigorous, and quantitative verification and validation (V & V) efforts, including targeted demonstrations. Though beyond the scope of this paper, a comprehensive overview of V & V challenges and progress as related to ICME has been captured in a paper by Cowles, Backman, and Dutton [6]. To establish confidence in application of ICME at any system level, an appropriate V & V plan must be established

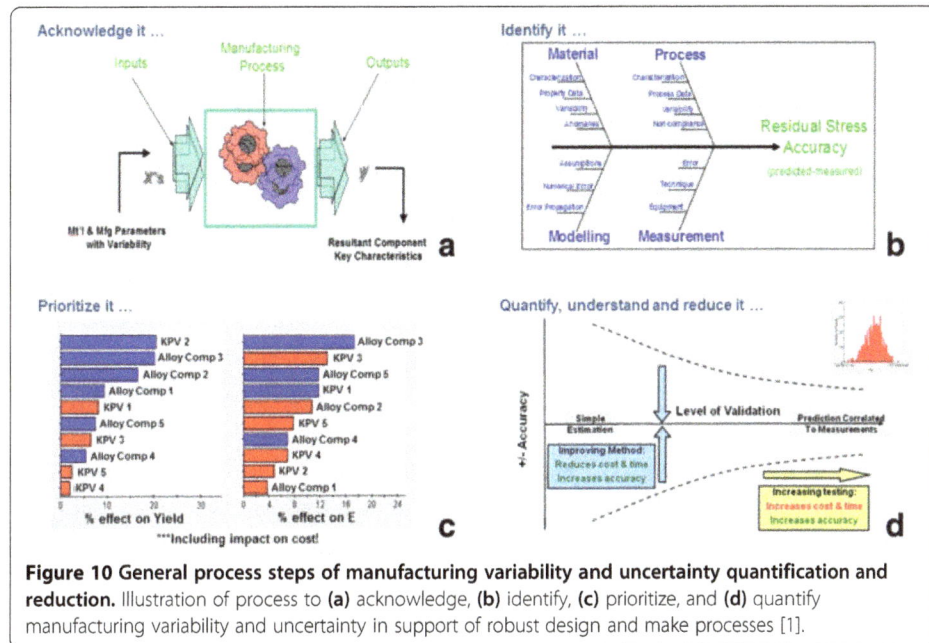

Figure 10 General process steps of manufacturing variability and uncertainty quantification and reduction. Illustration of process to **(a)** acknowledge, **(b)** identify, **(c)** prioritize, and **(d)** quantify manufacturing variability and uncertainty in support of robust design and make processes [1].

and executed to ensure that the modeling methods have been vetted to the level of accuracy required for the target application.

Use it

To provide specific context for integration of ICME into product development systems, a survey of the opportunities and challenges for use of ICME at different stages in the life cycle will now be reviewed.

Lifecycle integration examples and challenges

Preliminary design

Preliminary design occurs during phases 0 and 1 of the six-stage product life cycle (Figure 7). Stage 0 can occur over an extended period of time as the product requirements and the product attributes are developed simultaneously. In stage 0, product attributes of interest for gas turbine engines are performance, weight, cost, and life. When working closely with a customer, the two questions 'What can you do for me?' and 'What do you need from me?' are answered.

It is in this stage that the required material properties for a product are identified. ICME becomes integral to the product during this stage of the product life cycle. For example, the differing material properties for the rim of a turbine disc compared to the bore of the same disc are identified. The mathematical identification of these properties enables a robust design solution for product design.

Mathematical methods using design of experiments explore more solutions more thoroughly. The range of solutions establishes a design space where a product design can be selected that is less sensitive to variations. The design solution evolves from a top level solution where the design is established as an overall architecture, to detailed designs of sub-systems and components. This stair-step approach makes for not only effective product design by exploring the design space but also efficient solutions by keeping the solution from going off track. To include detail modeling of each component at the architectural selection phase would create a model too large to run on most computers and take too much time to set up such a model. The human interaction with the models between the levels of optimization needs to be effective in keeping the solution process on track toward a truly optimal product design.

The design of a component is a combination of the geometrical features and the material capability. Prior to ICME, the design of material capability was limited to selection of a material from a database with experimentally measured and validated nominal properties. ICME dramatically improves the design of a component by permitting the simultaneous detail design of geometrical and material features (microstructure and properties). By adding material features to the design space, the robustness of the product design is improved with the intent of reducing the chance of in-service issues. Figure 11 illustrates the cost and time benefits of getting a design right at the analysis phase.

Robust design requires mathematical representation of material properties as well as geometrical features and applied loads. ICME provides not only material capability in mathematical terms but also delivers location-specific properties that can be used to design and produce a product with superior attributes.

Rolls-Royce has been effective in performing robust design using Isight™ as the integrating framework for a number of programs as illustrated in Figure 12. From the first performance models, Rolls-Royce includes requirements for modules for engine modules and components as well as aircraft-required attributes. The outcome is an optimized, robust design. Numerous solutions are studied with varying requirements and defined optimizations to understand the design space. As the customer and Rolls-Royce gain confidence in the product strategy, the design iterations become more focused on the details of the modules then the details of the design. ICME is integrated with the design approach; first by various validated design allowables early in the process then by detailed process models. The result is an optimized product design.

Materials design

Feasibility evaluation, development, and insertion of new and optimized materials technologies have become heavily dependent on the use of ICME tools and

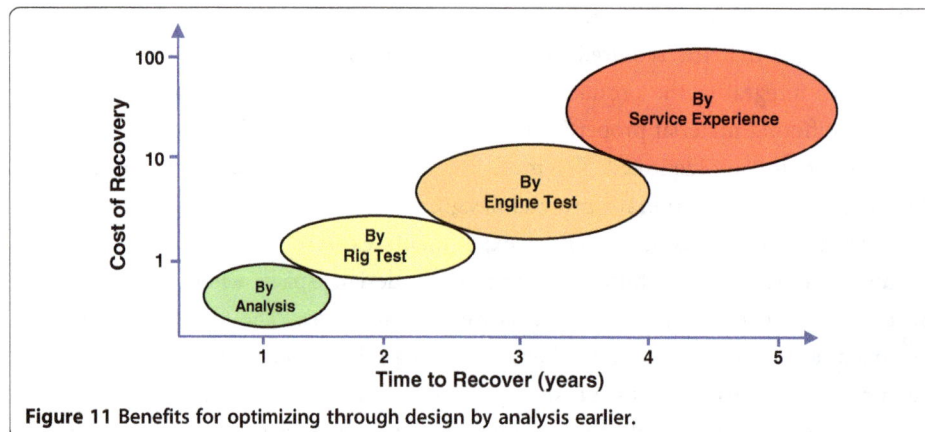

Figure 11 Benefits for optimizing through design by analysis earlier.

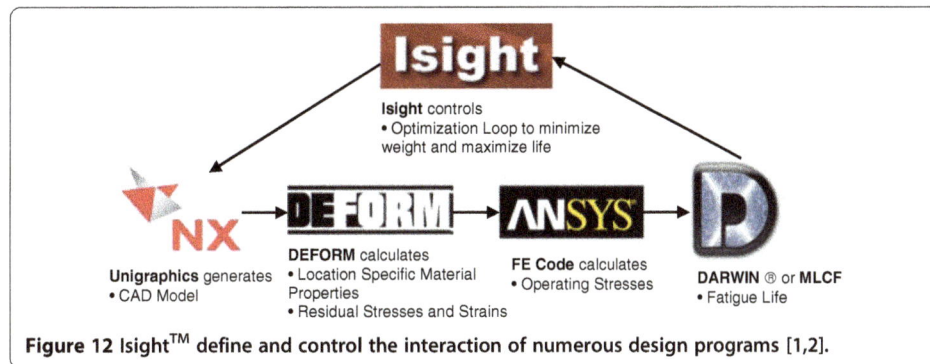

Figure 12 Isight^{TM} define and control the interaction of numerous design programs [1,2].

methods, as experimentation and empirical approaches alone cannot keep pace with current advanced aerospace component design cycles. The development and growth of computational tools for materials engineering have historically been hampered by the complexity and diversity of phenomena and properties that must be captured in order to deliver a robust and integrated engineering solution. Fortunately, ICME tools and the computing power required to process complex and simultaneous events during manufacturing and in service are now reaching a level of maturity where they can have a substantial impact if they can be integrated into product development [7].

A common and successful application for ICME tools has been the prediction of multicomponent alloy solidification behavior and phase equilibria in order to more efficiently assess novel compositions for structural materials. For example, innovations in existing nickel-based superalloy compositions and processing routes are required in order to overcome the challenges of usage at higher temperatures in advanced and future aerospace gas turbine engines. The most significant materials limitations for advanced γ/γ' (γ'') alloys in particular are phase instability above 700°C and mechanical strength reduction from segregation-induced defects. Eutectic γ/γ'-δ alloys (Figure 13) may offer a novel solution due to the additional reinforcing phase δ, which has demonstrated stability in experimental studies up to 1,000°C [8].

The rapid assessment of these potential next-generation alloy concepts has been made possible by the use of ICME tools. A commercially available software, with the incorporated Ni-alloy thermodynamic database, has demonstrated the ability to predict

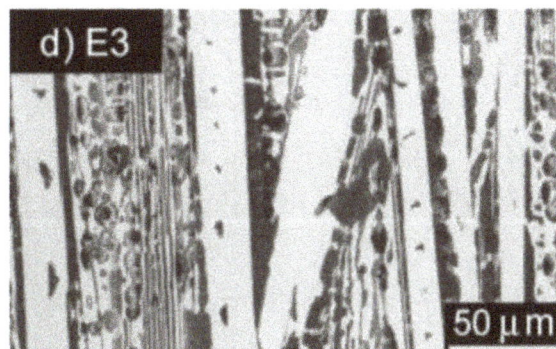

Figure 13 Representative micrograph of a γ/γ'-δ eutectic alloy, showing δ-phase plates within a γ/γ' matrix [8].

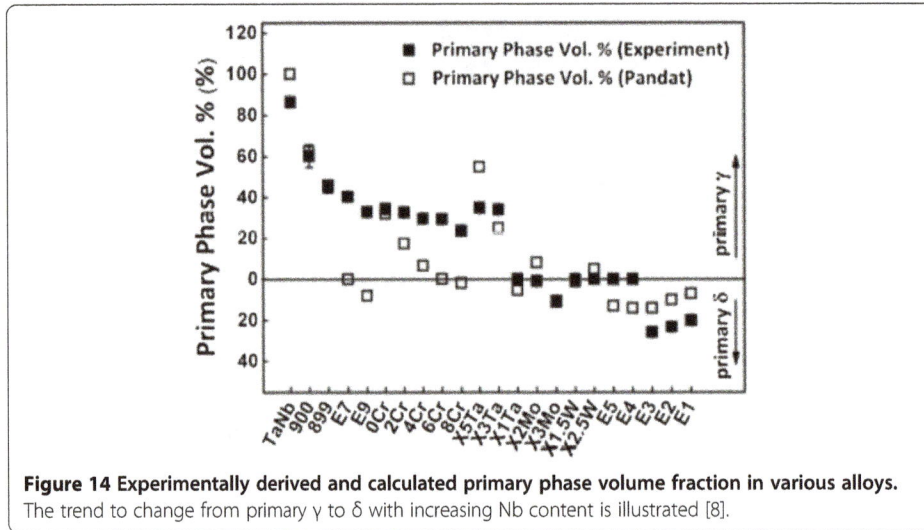

Figure 14 Experimentally derived and calculated primary phase volume fraction in various alloys. The trend to change from primary γ to δ with increasing Nb content is illustrated [8].

and confirm trends in primary phase volume fraction (Figure 14) and to confirm qualitative trends for phase equilibria and solidification behavior resulting from composition changes in multicomponent γ/γ'-δ alloys (Figure 15) [9]. Comparison to experimentally derived results has shown good agreement in some instances, whereas areas of disagreement between the experimental and predicted metallurgical data highlight gaps in the current thermodynamic database that require further optimization. Even given these current limitations for predictive capability, these particular ICME tools have resulted to a better understanding of the reasonable design space for a ternary γ/γ'-δ alloy system and provides the foundation for future exploration of these novel structural materials.

Detailed design

In today's highly competitive gas turbine industry, robust multidisciplinary design tools are required early in the design process which links several analyses to achieve aggressive overall system goals. An example of this detailed design process for a critical

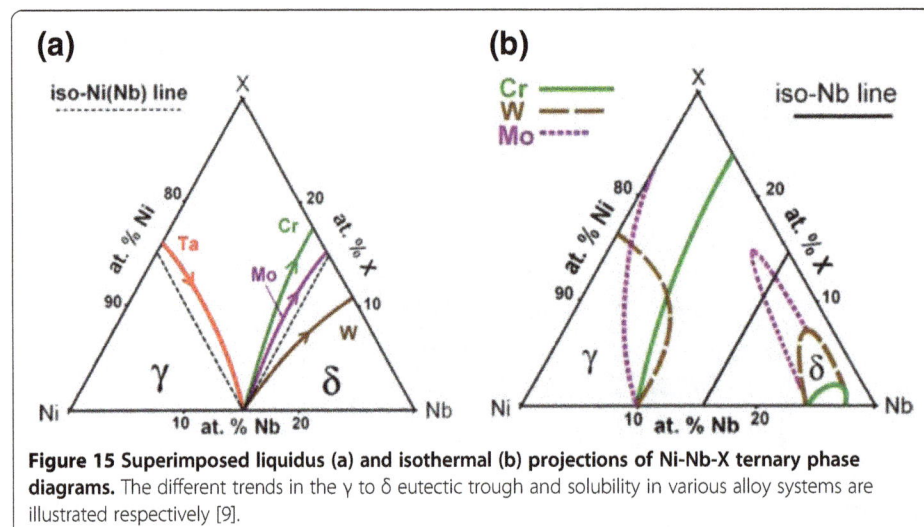

Figure 15 Superimposed liquidus (a) and isothermal (b) projections of Ni-Nb-X ternary phase diagrams. The different trends in the γ to δ eutectic trough and solubility in various alloy systems are illustrated respectively [9].

rotating component is shown in Figure 16. ICME is a relatively new addition to this design iteration loop and is highlighted in orange. A robust tool for sizing a new rotor disc would traditionally link aerodynamic, mechanical, and fluid systems design trades to the resultant heat transfer analysis, structural analysis, and predicted service life of the component. In this manner, design trades are quickly assessed, and the impact to the sub-system is quantified. Linking steps in the analysis allows perturbation of the system by making, for example, a change to a secondary fluid cooling scheme. Such a change would alter the predicted disc temperatures, resultant stress, and life prediction as well as change in blade tip clearance. Linking these types of tools can provide the design team an updated performance, life, and cost prediction indication in minutes.

With the advent of high-performance computing, linking these design tools becomes more attractive and affordable. Recent advances in materials modeling have resulted in analytical tools that package nicely into the design loop described previously. Combining with the traditional design loop, high-performance computing with quantitative material property and residual stress modeling becomes a powerful toolset for advancing the state of the art. Figure 16 shows the ICME modeling capability which are now included in the detailed design process. This approach was recently applied to a legacy LCF spin rig disc experiment. In this historical spin rig study, different initial residual stress profiles were imparted to discs by applying varying degrees of a single pre-spin event prior to cyclic testing. These pre-spin events created regions of local yielding, leading to localized compressive residual stress. Components were cyclically tested to crack initiation and fracture with and without beneficial residual stresses imparted due to pre-spin at critical locations. A 2.8× increase in total life to failure (crack initiation and crack growth) was observed for the pre-spun components compared to the baseline condition. Further details of this experiment are provided in depth by Shen et al.

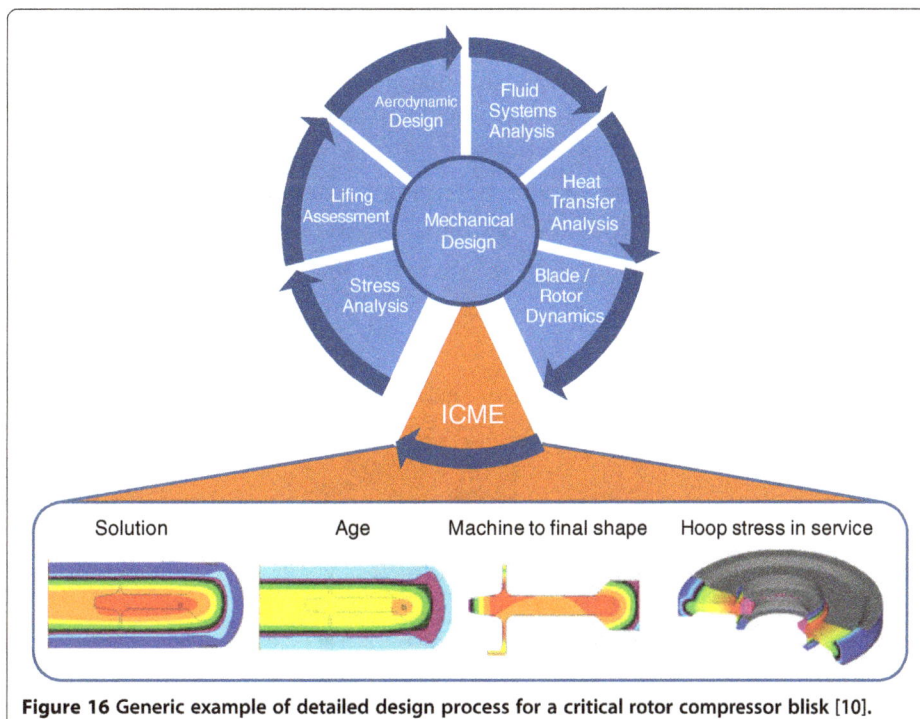

Figure 16 Generic example of detailed design process for a critical rotor compressor blisk [10].

[10]. Application of ICME toolsets was used to predict the component residual stresses by modeling the process of forging, machining, and pre-spinning operations performed prior to testing. Using inelastic stress calculations, the improved total life to failure observed in the spin test study could be predicted through the Rolls-Royce analytical lifing methodologies. Based upon the validation testing conducted previously on Rolls-Royce historical rigs, it has been shown that the accuracy of total life calculations can be improved by accurate account of induced residual stresses. By applying ICME toolsets, the residual stresses that form during the forging and heat treatment process can also be derived for a more accurate description and analysis of the finished component. These calculated residual stresses can then be combined with thermo-mechanical mission stresses for analytical low-cycle fatigue initiation and fatigue crack growth life calculations. Successful control of the manufacturing process allows for tailoring of the final component residual stress and potential for total life improvement in low-cycle fatigue crack initiation and fatigue crack growth.

The work done by Shen et al. [10] opens up the possibility of tailoring the forging shape and/or heat treatment cooling rates to achieve an optimized component solution. Such solution may be minimum weight - a full life component design that meets the sub-system goals. Figure 17 shows the disc used in the previously described spin test with two sample forging shapes. This figure illustrates the variables that can be controlled to tailor the material properties and residual stress of the finished part. Namely, the local heat transfer coefficient (as a function of time) and forging geometry are the key process variables contributing to the resultant predicted service life. This becomes an ideal problem for DOE and/or optimization techniques. Rolls-Royce has recently conducted this type of work for several critical components in recent advanced system demonstrators to maximize service life. This required close working with suppliers to achieve an overall acceptable solution for the program. The result was a specific forging geometry and heat treat fixture that provided acceptable material properties and desired initial residual stress profile. In addition, predicted deflections, obtained by simulating the machining process, were very close to what were observed in manufacturing. Predicted deflections from this modeling were used to guide manufacturing to deliver a complicated component to print without deviation.

The introduction of ICME techniques to the design of gas turbine engines presents a potentially game-changing technique for designing critical engine components for higher life. By considering the residual stress inherent to any forged and heat-treated component allows the designer to make more efficient use of material, weight, life

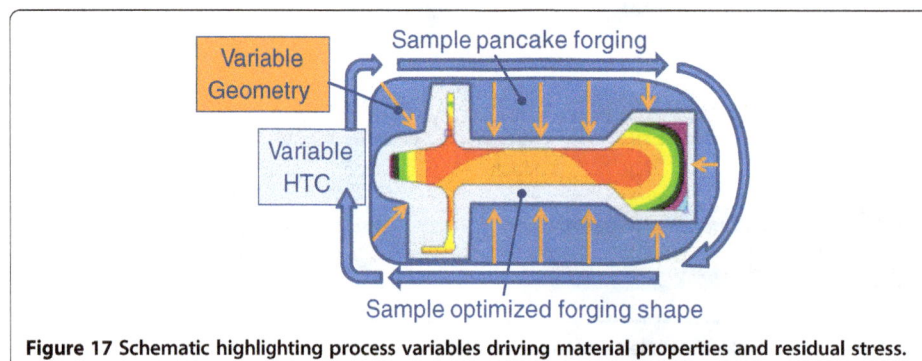

Figure 17 Schematic highlighting process variables driving material properties and residual stress.

extension, and/or cost design trades. For existing products, this approach can be applied to extend the life of fielded components through alterations to the heat treatment process or forging geometry only, limiting the cost impact to implement changes. Such changes are often warranted later in the design life cycle as engineers gain confidence in how components are actually used and maintained in service. In addition, understanding of the inherent manufacturing process variability can produce a more robust assessment of fielded component risk. This increased understanding can be used by the fleet operators to improve decision making on maintenance and overhaul scheduling in addition to reliability predictions. The benefits of realizing and incorporating this technology are just now beginning to come to fruition. Rolls-Royce is now updating their design methodology to include material and process modeling in the design process. While much work remains to fully benefit from the potentials of including ICME into our design practices, initial work has shown positive trends in capturing additional life and minimizing part weight.

Virtual engine

Detailed design and analysis of individual components and sub-systems alone are not sufficient to fully realize/maximize aero engine performance. Improved design understanding of the entire system is required in order to optimize the final product according to environmental and operational constraints. Significant recent advances in high-performance computational facilities enable analysis of many sub-systems in one 'product level' model with sufficient geometric detail and boundary condition definition required to simulate system behavior, such as component movements under steady-state and transient loads. Many of today's design tools do not sufficiently account for material and manufacturing variability and their impact on component and assembly shape, structural integrity, and performance. For gas turbine design, system level 'virtual engine' models incorporating multidisciplinary analysis (i.e., aerodynamics, mechanics, materials, and performance) are giving greater insight to the impact of system variations on overall system performance, cost, weight, and life (Figure 18) [11].

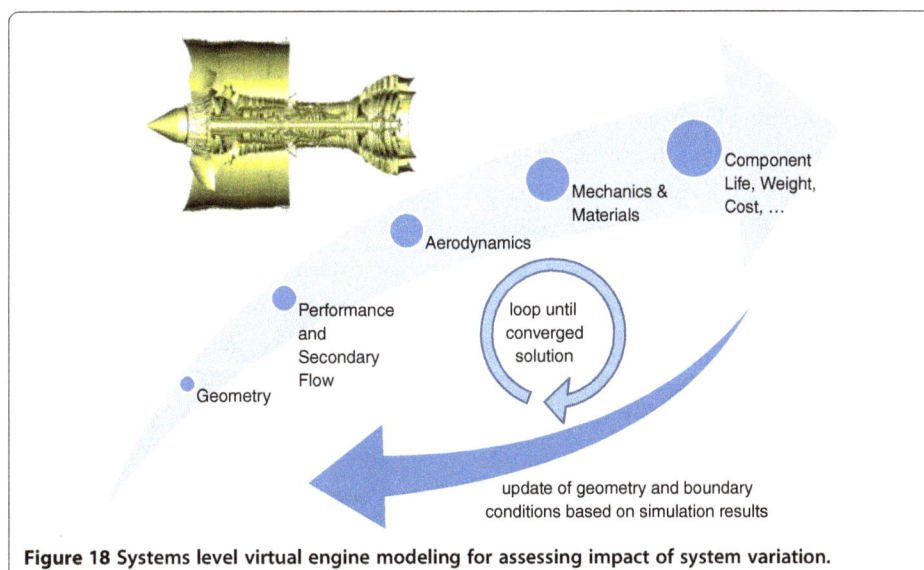

Figure 18 Systems level virtual engine modeling for assessing impact of system variation.

Virtual manufacturing

Manufacturing is traditionally treated as an entirely separate function within the life cycle of a product. *Design engineering* will perform studies and develop various iterations of a product design then deliver a final definition to *manufacturing*. This definition may be in drawing or electronic formats. *Manufacturing engineers* will then determine a method to manufacture the product. In case of a new product design for a machined component, a manufacturing engineer will define a sequence of operations utilizing machine tools, fixturing, cutting tools, and inspection equipment based on available experience and knowledge. Once a method is defined, tooling/fixturing has been designed and built, and NC programs have been written, the process of prove out begins. Each step in the method will be proven out on the production machine tool. If any step in the method does not perform as planned, then modifications are made to the process and tried again. This can become a lengthy and costly loop, especially when there are numerous machining operations. Inherent variations within the process can make troubleshooting of issues very difficult and time consuming. During these prove-out steps, production machine tools are being utilized for development and not for daily production. This reduces operational productivity, disrupts material flow, complicates scheduling and could lead to late deliveries. Furthermore, delay of validation testing of components (required to be conformed to final production processes) may also result in certification delays and additional costs.

The integration of technologies like ICME and model-based definition (MBD) into the product development process facilitates digital integration of manufacturing knowledge, standards, methods, capabilities, and limitations into the design development process. This allows engineering 'design and make' teams to predict, optimize, and prove out a components method of manufacture before any step of the manufacturing process is started. The condition of supply definitions can be optimized to minimize variations within a chosen method of machining sequences. Fixturing can be designed to provide the best holding capability and minimized part distortion. NC programs and cutting strategies can be fine-tuned for performance, capability, and robustness. All of these steps can go through numerous iterations virtually to achieve the most optimized process that will perform as predicted, and much of it can be done even before the design is finalized. When fully mature, this approach allows for a new component to be introduced into a production line Right First Time with a process performing at, or better than, the capability of existing processes.

To fully realize this vision, however, integrated tools that span the full supply chain are needed to achieve full capability. Robust modeling tools should be integrated with the Computer Aided Manufacturing systems to allow for full material characteristics prediction and optimization. Challenges arise when an external supply chain is involved, however, since systems that are internal to a company are much more straightforwardly integrated. The overall supply chain needs the ability to integrate their systems and processes to enable true lifecycle integration. IT security, export control and compatibility issues all add to the complexity of a global 'digital' supply chain.

Virtual simulation of the machining process exists today in most of the modern CAD/CAM systems and external applications such as Vericut™ by CGTech; however, most current manufacturing process simulations do not integrate available ICME toolsets for prediction of manufacturing process impact on part yield, material properties and/or design performance. Vericut™ for example can simulate the complete kinematics motion of a multiaxis machining center and interaction of the cutting tools and fixtures. Vericut™ can also simulate material removal by cutting tools and compare the cutter path resultant to the original part model. These simulations are valuable for a virtual prove out of the cutter path accuracy and potential interferences between tools/fixtures/machines/parts; however, these simulations are static and will not take into account cutter deflections or part movement from relieving residual stresses as material is cut away. To realize the full potential of an integrated virtual manufacturing system, ability to predict and simulate how the materials interact and react during an operation or process is critical. Fully integrated virtual manufacturing systems could eliminate the costly physical prove-out stages in manufacturing and facilitate Right First Time solutions when integrating known process capability to design-for-manufacture toolsets earlier in the product development process. New products could be introduced into production along with existing products, allowing manufacturing to improve management inventory while maintaining the 'heartbeat' of the product flow through the physical manufacturing facility itself.

As integration of virtual manufacturing across the supply chain matures, virtual manufacturing systems at the component level are already delivering substantial business benefit in specific cases. One example is in the virtual modeling of the casting process (Figure 19). Historically, casting process development has been performed through an expensive 'build and break' approach [12]. Through the use of casting modeling during product and process development, engineers can evaluate new designs for manufacturability and optimize the manufacturing process before real components are

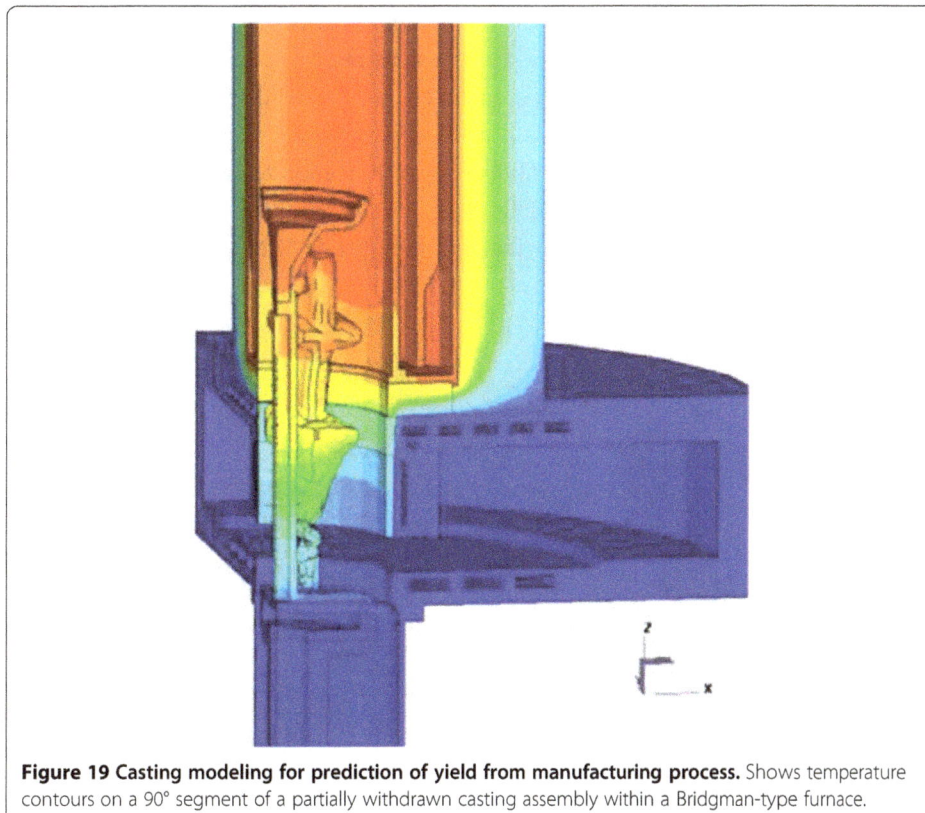

Figure 19 Casting modeling for prediction of yield from manufacturing process. Shows temperature contours on a 90° segment of a partially withdrawn casting assembly within a Bridgman-type furnace.

cast, during new product introduction or yield improvement activities. By accurately modeling the heat exchange between the casting, mold, and key parts of the furnace, Rolls-Royce has been able to successfully predict porosity, stray grains, and freckle chain defects. Improvements proposed and rapidly tested by iterating the model can then be validated by selected and highly targeted physical trials, significantly reducing the cost and lead time associated with build-and-break-type process development efforts. Investment casting assemblies now typically contain features that would likely not have been developed without the use of modeling. Rolls-Royce has realized significant business benefit from deployment of process modeling systems to casting facilities and suppliers, which facilitate the use of ICME tools by engineers without requiring a high level of ICME knowledge or skill. Such deployed application of modeling has been applied to modify tooling, process, and product designs to deliver significant cost reduction through reduction of non-conformance and improvements in Right First Time production.

Life prediction and management for life-limited components

While ICME offers powerful new predictive tools, it also embraces the need to quantify uncertainty in predictive capability as seen in the preceding examples. It follows, from a business perspective, the NRC's vision of ICME as a holistic system that brings together design optimization and manufacturing that encourages further consideration of uncertainties throughout the entire life cycle of a product, not just during development. This is partly due to the nature of optimization and partly to do with the role manufacturing plays in life cycle cost reduction (see Figure 20).

While the objective functions for system design optimization are generally well understood, this is not always true for system boundary conditions. Often, engineers' understanding of boundary conditions improves as more data is acquired from manufacturing and the product's usage in service. For the same reason, it may become apparent as a product matures that it satisfies intended requirements but is no longer optimized for the conditions to which it is actually exposed. Further, as such information becomes apparent, tools like ICME will likely be called upon to help improve performance and cost. There are many possible paths toward product improvement and cost reduction to explore. Early use of engineering expertise in ICME along with later

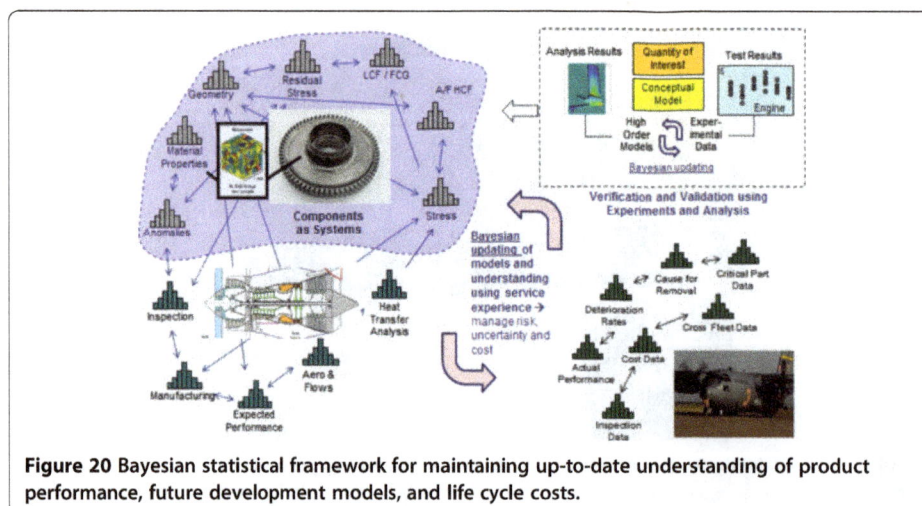

Figure 20 Bayesian statistical framework for maintaining up-to-date understanding of product performance, future development models, and life cycle costs.

in-service data to update models and prior understanding means that ICME fits well within a broader context of Bayesian frameworks and UQ-based decision analysis [13,14]. Perhaps one of the best examples in the gas turbine industry is in life prediction and management of high-energy discs.

In application to high-energy discs, computational materials science tools must be carefully linked with many other UQ and decision-making tools and data sources to enable the NRC vision [7]. Design optimization and field management of these components require application of a number of complex, inter-related tools, technologies, physics-based models, cost and reliability models, as well as expert decisions. Moreover, referring back to Figure 9, there are many sources of uncertainty associated with discs. Some of these can be quantified early by ICME's development models such as microstructural characterization, forging, and finished part residual stresses and material properties. Other uncertainties such as operational usage and typical engine performance and deterioration effects must be assumed prior to certification and production. However, in both cases, what are commonly known as Bayesian prior distributions must be assumed. In the case of usage, distributions associated with environment such as ambient conditions or distributions reflecting uncertainty in operator behavior such as pilot decisions to use lower power flex takeoff conditions may be defined. In the case of material uncertainties, the *supply chain* processing parameters and manufacturing target dimensions may be considered Bayesian priors. Furthermore, these two general sources of uncertainty, developmental and in-service, must be updated and managed jointly in order for safety, performance risks, and costs to be effectively managed throughout the life cycle of the product. This can easily be described by way of a simple example if one considers the correlated effects of inelastic material deformation from speed and temperature with location-specific material characteristics such as yield strength. Bayesian posterior distributions, e.g., LCF life, will therefore be functions of *both* material properties and *usage*. As described earlier, ICME then becomes a key part of continual validation and verification. Measureable data such as forging solution temperature, distortion from machining, finished part geometry, and engine air temperatures can be used to update corresponding distributions of quantities that are more difficult to measure in production such as component residual stresses, metal temperatures, and disc stresses. This process of Bayesian updating can be used to generate highly correlated distributions such as disc creep life and damage tolerance capability that naturally follow out of similar prior understandings and data. Furthermore, UQ tools that lend themselves to quantifying system behavior such as *Bayesian networks* can be used to maintain an updated understanding of these distributions [14]. For discs (and other components), these tools are envisioned to work alongside standard statistical and *robust design* methods to link ICME into a holistic production system.

Many UQ tools themselves, like Bayesian networks, will be largely invisible to the supply chain. But the outputs, questions, and decisions associated with these networks will be immediately visible and will be easy to explain in terms both an OEM and supplier will understand. In this way, both parties can supply and discuss implications to both non-proprietary and proprietary models and information from new data as it becomes available. The basic Bayesian reasoning process includes simple cause and effect relationships, prior distributions from expert knowledge and data, likelihood models, and approach to updating understanding. This makes it a natural tool for engineers developing complex systems in the context of ICME. ICME models are envisioned as a powerful new way to link design and the manufacturing supply chain. But as with any model, its ultimate utility is seen in how it is used by those who make business risk-related decisions.

Conclusions

Significant work has been done across the industry to develop computational models of materials, manufacturing, and design processes; however, there has been less success at integrating these models into product development processes. Application of these models has delivered both development and production benefits, but mostly in isolated one-off situations. The integration of ICME into design and manufacturing systems, though clearly more challenging, promises order of magnitude benefits in cost and lead time reduction. This review has illustrated and emphasized that integration challenges are not just digital but physical, heuristic, and relational as well. To be successful, ICME initiatives must deliver solutions the entire supply chain understands, trusts, and can use. A review of some of the specific challenges faced by ICME users across the supply chain was presented. Ultimately, the transition from technology push to technology pull for use of ICME across the supply chain will reside in the benefit realized through successful use of the technology to deliver product benefit. In response to already realized business benefit on both legacy and new product introduction, Rolls-Royce is integrating ICME into our product development processes as part of our standard methodology going forward.

Competing interests
The authors declare that they have no competing interests.

Authors' contributions
JM conceived, organized, and integrated all individual author contributions to draft the review. AB contributed to materials design content. PB, DH, and KM contributed to virtual manufacturing content. RC contributed to preliminary design content. NC contributed to detailed design content. CD contributed to the detailed design and life prediction and management content. AK contributed to virtual engine content. GM contributed to uncertainty quantification and robust design content. JS contributed to uncertainty quantification, Bayesian methodology and life prediction and management content. All authors read and approved the final manuscript.

Acknowledgements
Authors would like to express appreciation to Dr. Rollie Dutton and Dr. David Furrer who initially invited a presentation of this work under the same title which was presented during a symposia on 'ICME - from the customer's view point' that Dr. Dutton and Dr. Furrer co-chaired during the 2012 Materials Science & Technology conference in Pittsburgh, PA. Authors would also like to acknowledge contributions from Ben Saunders of Rolls-Royce who supported with images for figure generation and Tony Phipps of Rolls-Royce who supported business benefit articulation. Finally, it is prudent to recognize the purposeful integration efforts being coordinated across several professional societies including TMS, ASM International and AIAA. The lessons learned reported in this paper are collated in part from cross-industry and academia collaboration and feedback regarding the ICME integration efforts of these communities.

Author details
[1]Rolls-Royce Corporation, Indianapolis, IN 46225, USA. [2]Rolls-Royce PLC, Derby DE24 8BJ, UK.

References
1. Matlik JF, Bolcavage A (2012) Integrated computational materials engineering from a gas turbine engine perspective. In: Symposium of integrated computational engineering - the customer's point of view. Presentation at the Materials Science & Technology 2012 Conference & Exhibition, Pittsburgh, PA. 8 October 2012
2. Glavicic MG, Goetz RL, Shen G, Rasche JA, Ress RA, III, Stillinger J (2011) Application of ICME to turbine engine component design optimization. In: 52nd AIAA/ASME/ASCE/AHS/ASC structures, structural dynamics and materials conference. Denver. 4–7 April 2011, AIAA-2011-1738
3. National Research Council (2004) Accelerating technology transition: bridging the valley of death for materials and processes in defense systems. The National Academies Press, Washington, DC

4.   Olson GB (2013) Integrated Computational Materials Design: From Genome to Flight. In: 54th AIAA/ASME/ASCE/
     AHS/ASC structures, structural dynamics and materials conference. Boston. 8–11 April 2013, AIAA-2013-1847
5.   Martin R, Evans D (2000) Reducing costs in aircraft: the metals affordability initiative consortium. JOM 52(3):24–28
6.   Cowles B, Backman B, Dutton R (2012) Verification and validation of ICME methods and models for aerospace
     applications. Integr Mater Manuf Innov 1:2
7.   National Research Council (2008) Integrated computational materials engineering: a transformational discipline for
     improved competitiveness and national security. The National Academies Press, Washington, DC
8.   Huron ES, Reed RC, Hardy MC, Mills MJ, Montero RE, Portella PD, Jack T (2012) Polycrystalline γ (Ni)/γ' (Ni$_3$A1)-δ
     (Ni$_3$Nb) eutectic Ni-base superalloys: chemistry, solidification and microstructure. In: Xie M, Helmink RC, Tin S (ed)
     Superalloys. Wiley, Hoboken, NJ, pp 633–642
9.   Villars P, Okamoto H, Cenzual K (ed) (2007) ASM alloy. ASM International, Phase Diagrams Center, Materials Park, OH
10. Shen G, Cooper N, Ottow N, Goetz R, Matlik J (2012) Integration and automation of residual stress and service
     stress modeling for superalloy component design. In: Huron ES, Reed RC, Hardy MC, Mills MJ, Montero RE, Portella
     PD, Telesman J (ed) Superalloys 2012. Wiley, Hoboken, pp 129–134
11. Keskin A, Saiz A (2012) An integrated design system for optimization of gas turbine components. In: 53rd AIAA/
     ASME/ASCE/AHS/ASC structures, structural dynamics and materials conference. Honolulu. 23–26 April 2012,
     AIAA-2013-1408
12. EuroPAM (2006) 16th European conference and exhibition on digital simulation for virtual engineering. Toulouse.
     10–12 October 2006
13. Azaidan M, Mills A, Harrison R (2013) Bayesian framework for aerospace gas turbine engine prognostics.
     In: IEEE aerospace conference. Big Sky. 2–9 March 2013
14. Mahoney S, Stillinger J (2012) Towards a probabilistic framework for integrated computational materials
     engineering. in: 53rd AIAA/ASME/ASCE/AHS/ASC structures, structural dynamics and materials conference.
     Honolulu. 23–26 April 2012. AIAA-2013-1529

3

Making materials science and engineering data more valuable research products

Charles H Ward[1*], James A Warren[2] and Robert J Hanisch[2,3]

* Correspondence:
charles.ward.4@us.af.mil
[1]Air Force Research Laboratory,
Materials and Manufacturing
Directorate, Wright-Patterson AFB,
OH 45433, USA
Full list of author information is
available at the end of the article

Abstract

Both the global research community and federal governments are embracing a move toward more open sharing of the products of research. Historically, the primary product of research has been peer-reviewed journal articles and published technical reports. However, advances in information technology, new 'open access' business models, and government policies are working to make publications and supporting materials much more accessible to the general public. These same drivers are blurring the distinction between the data generated through the course of research and the associated publications. These developments have the potential to significantly enhance the value of both publications and supporting digital research data, turning them into valuable assets that can be shared and reused by other researchers. The confluence of these shifts in the research landscape leads one to the conclusion that technical publications and their supporting research data must be bound together in a rational fashion. However, bringing these two research products together will require the establishment of new policies and a supporting data infrastructure that have essentially no precedent in the materials community, and indeed, are stressing many other fields of research. This document raises the key issues that must be addressed in developing these policies and infrastructure and suggests a path forward in creating the solutions.

Keywords: Materials data; Data policy; Data repository; ICME; MGI; Integrated Computational Materials Engineering; Materials Genome Initiative; Data archiving

Introduction

Reliance on shared digital data in scientific and engineering pursuits - whether the data are derived from computation or experiment - is becoming more commonplace within the materials science and engineering (MSE) community. Concurrently, government policies across the globe are embracing an 'open science' model which sets a requirement for sharing digital data generated from publicly funded research. A recent joint Materials Research Society and The Minerals, Metals and Materials Society (MRS-TMS) survey on 'big data' in materials science and engineering showed that 74% of respondents would be willing to participate in sharing their data if it was encouraged as a term and condition of funding or publishing, assuming the proper safeguards were in place [1]. However, it is fair to say that the MSE community currently lacks the strategy, framework, standards, and culture needed to support materials data curation and sharing. A unified approach is needed to meet the growing demands of the community and a plan to meet government requirements for broad access to digital data. It is clear that the peer-reviewed journals and government-sponsored technical reports serving the MSE community can be an essential component to the solution, and there is now an opportunity to proactively plan how they may best serve the growing needs of their constituency.

We have structured this paper to, first, provide the reader a general awareness of the global environment and ongoing activities concerning the management of research data. We then present a perspective on the benefits of data archiving to the MSE community and outline what attributes and characteristics a data archiving solution should have. We follow this with a discussion of key challenges yet facing the establishment of a digital materials data infrastructure. Finally, we propose a way forward to tackle the creation of community-based solutions for data archiving policies and data repositories.

Review

Global context

The 2008 NRC report on Integrated Computational Materials Engineering (ICME) highlighted the importance digital data will play in the future of materials science and engineering [2]. MSE's ever increasing reliance on computational modeling and simulation will demand digital data as the feedstock for solutions in both science and engineering.

In the USA, the National Institutes of Health have long promoted a policy of open access to data generated from their grants [3]. In the mid-1990s, the Human Genome Initiative spawned the Bermuda Principles which called for immediate public posting of sequences of the human genome [4]. More recently, the National Science Foundation has adopted a requirement that applicants provide a data management plan in grant proposals [5]. Specific to the materials community, the sharing of digital data is a key strategy component of the US's Materials Genome Initiative (MGI), and mechanisms to foster and enable sharing are actively under consideration [6].

The European Union has been very proactive in studying the impacts of a digitally linked world on the scientific community. The EU Framework Programme 7 funded a project called Opportunities for Data Exchange that has produced several relevant reports on publishing digital data in the scientific community [7]. And, in June 2012, the Royal Society published 'Science as an open enterprise' which promotes free and open access to scientific results, including data [8]. These studies are now broadly informing government policy. For example, recent policy issued in the UK in July 2012 calls for government-funded research to be published in open access journals and requires access to supporting research data [9]. In February 2013, Dr. John Holdren, director of the Office of Science and Technology Policy (OSTP), issued a directive to all federal agencies to develop plans to make the results of federally funded research more accessible to the public [10]. A key component of this directive is a call for agency plans to include a means by which digital data resulting from research can be made available to the public. In support of this policy, the White House has established a useful web site providing resources supporting the establishment of open data [11]. The US Government funding agencies have since provided their plans to address OSTP's open research policy, and results are imminent.

Other technical communities have addressed the challenges of access to digital data with a variety of approaches. Indeed, the biology community has implemented a number of differing approaches; for example, the approach taken in genetics versus that adopted by evolutionary biology [12,13]. In other disciplines, one subfield of thermodynamics has already adopted a very structured approach to archiving data, while the earth sciences community has embarked on an effort to define its approach [14,15]. The astronomy community has dedicated international resources to the development of the Virtual Observatory, an infrastructure that enables global data discovery and access across hundreds of distributed archives [16]. Despite the differing mechanics of implementation, all the approaches were rooted in a community-led effort to define the path best suited for that particular technical field.

In response to these trends, technical communities and publishers have developed and implemented open access journals and data archiving policies. Again, the field of biology appears to be leading the way on both these fronts. One example of this trend is *Database: The Journal of Biological Databases and Curation*, an open access journal dedicated to the discussion of digital data in biology [17]. And in a recent development, Nature Publishing Group has launched a new open access journal, titled *Scientific Data*, which is dedicated to publishing descriptions of scientific datasets and their acquisition [18]. It will initially focus on life, biomedical, and environmental science communities. Furthermore, the Public Library of Science recently strengthened its policy on data access: 'PLoS journals require authors to make all data underlying the findings described in their manuscript fully available without restriction, with rare exception' [19].

In order to begin a dialog within the MSE community, the National Institute of Standards and Technology (NIST) convened a workshop on digital materials data in May of 2012 under the auspices of MGI. The workshop identified a number of barriers that need to be addressed during creation of a data strategy for materials, they include: materials schema/ontology, data and metadata standards, data repositories/archive, data quality, incentives for data sharing, intellectual property, and tools for finding data [20]. Other disciplines, notably evolutionary biology, have demonstrated that peer-reviewed journals have the potential to contribute solutions to these barriers to data sharing [21].

Benefits of archiving materials science and engineering data

There is a growing realization within the global scientific community that the data generated in the course of research is an oft overlooked asset with considerable residual value to other scientists and engineers and that often a significant portion of the data is stored but not accessible. The following are several anticipated benefits of increasing access to materials science and engineering data in digital form:

Data reuse

- Scientific productivity and return on investment in research infrastructure.
- Secondary hypothesis testing.
- Reducing/eliminating paying for data generation multiple times.
- Comparing with previous studies.
- Integrating with previous and future work.
- Reproducing and checking analyses.
- Simplifying and enhancing subsequent systematic reviews and meta-analyses.

- Facilitating interdisciplinary research.
- Teaching.

Incentives

- Increasing academic credit (citations).
- Access to one's own data at a future date.
- Convenience and security of cloud storage.

Others

- Testing algorithms/computations with validated reference datasets.
- Meeting funding agency requirements to share data.
- Reducing the potential for duplication of effort.
- Reducing of error and fraud.

The MRS-TMS 'big data' survey asked participants to evaluate whether given attributes would act as impediments or motivators to sharing data, Figure 1 [1]. The bottom of the graph shows that the largest impediments are primarily driven by legal considerations. The top of the graph demonstrates that the strongest positive motivators are the increased attention and credit a researcher may draw for one's work. It is clear that widespread sharing of digital materials data will require not only technological advances but also cultural shifts that include modernization of traditional incentives for the sharing of scholarly works to include recognition for publication of data. While there is no universally accepted solution available at present, new tools such as the Thomson Reuters Data Citation Index may help provide avenues for this recognition [22].

The impact on research productivity owing to the provision of well-calibrated, well-documented, archival data products is clearly demonstrated in the case of NASA's Hubble Space Telescope (HST). Initially, archival data was not used very extensively; the data suffered from spherical aberration, of course, resulting in a factor of approximately ten decrease in sensitivity from expectations. But in the early 1990s, there was

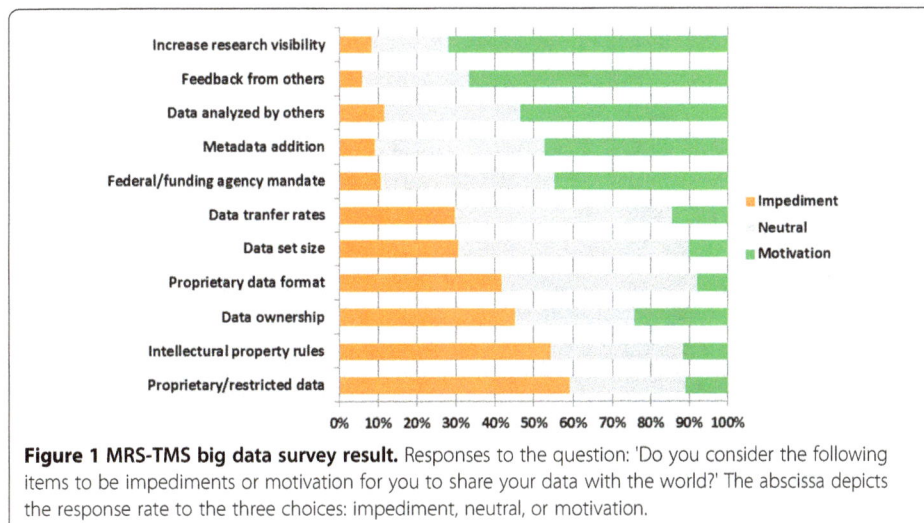

Figure 1 MRS-TMS big data survey result. Responses to the question: 'Do you consider the following items to be impediments or motivation for you to share your data with the world?' The abscissa depicts the response rate to the three choices: impediment, neutral, or motivation.

also somewhat of a stigma attached to using archival data for research: this was some-how not as good or pure as collecting one's own data at a telescope. But times have changed, and HST archival data is now used extensively by astronomers unaffiliated with the teams making the original observations. The resulting research papers ac-count for more than half of all peer-reviewed publications based on HST observa-tions (see Figure 2). There are a number of reasons for the big increase in archival data use. HST observing time is very difficult to get, with typically a seven-to-one oversubscription ratio in the proposal process. All HST data is routinely pipeline processed, yielding an archive of 'science ready' data products. All HST data be-comes public after a nominal 12-month embargo period. And HST data taken for one purpose can often be utilized for studies of a substantially different intent. While this high level of reuse may not be achieved for all research experiments, the HST example clearly shows that a substantial improvement in research prod-uctivity can be achieved, at a very modest incremental cost, when proper care is taken in designing the data management system.

   In materials science, there is a strong case that data obtained at great public ex-pense should be made available to as large a group of researchers as is practical, noting that, unlike astronomical data, there can be important constraints due to both national security and intellectual property concerns. Acknowledging these is-sues, however, materials data obtained from national user facilities, such as data obtained by scattering of synchrotron light or neutron beams, are examples of a scarce yet valuable information stream. Indeed, these facilities have recognized the importance of good archiving capability but have not focused on distribution of these data. Making such information more widely available would increase the amount of materials knowledge discovery with a small investment relative to the costs of repeating the work.

Background for data archiving

A common approach to archiving materials data has several benefits, but its primary value would be to provide unified, consistent guidance and expectations throughout the scientific and engineering community. However, while the development of an archiving policy itself may be relatively straightforward, the infrastructural issues

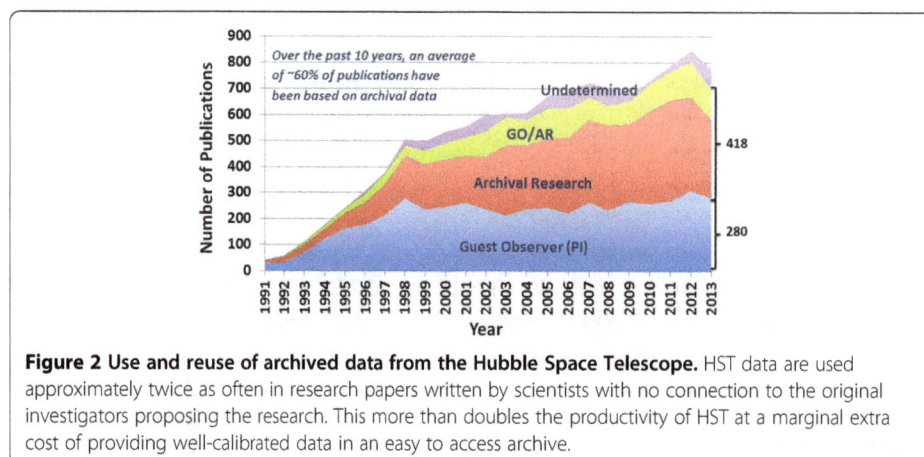

Figure 2 Use and reuse of archived data from the Hubble Space Telescope. HST data are used approximately twice as often in research papers written by scientists with no connection to the original investigators proposing the research. This more than doubles the productivity of HST at a marginal extra cost of providing well-calibrated data in an easy to access archive.

necessary to support policy implementation are extraordinarily complex. These issues include the establishment of viable

- Repositories for materials data.
- Standards for data exchange.
- Citation and attribution protocols.
- Data quality metrics.
- Intellectual property and liability determinations.

Characteristics of an archiving solution

In order for a data archiving solution to be of lasting value to researchers and maintain the rigorous, archival standards of relevant publications, it should have the following minimum set of characteristics:

- Persistent citation.
- Data discoverability.
- Open access (for journals).
- Ease of use.
- Minimal cost.

To archive or not to archive?

The most critical question to be answered in setting policies for publications is 'what data should be archived?' The answer is essential in providing clear expectations for authors, editors, and reviewers, as well as determining the size of the data repositories needed. Other disciplines have already embarked on this journey and have devised a variety of approaches that suit the data needs of their communities for their stage of 'digital maturity.' Two ends of the spectrum in addressing this question are presented here. The first assumes all data supporting a publication are worthy of archiving. This criterion is found most often in peer-reviewed journals that have narrow technical scope and generally deal with very limited data types. For example, journals in crystallography and fluid thermodynamics have very stringent data archiving policies that prescribe formats and specific repositories for the data submitted [23,24]. Other journals that cover broader technical scope, and therefore deal with more heterogeneous data, have implemented more subjective criteria for data archiving and a distributed repository philosophy. Earth sciences and evolutionary biology have typically taken this approach. It is likely that the approach adopted by MSE publications may also span a similar spectrum, depending on the scope of the publication.

The MRS-TMS 'big data' survey provided insight into the community's perspective on the relative value of access to various types of materials data, shown in Figure 3. It is interesting to note that as the complexity of the data and metadata increase (generally) toward the right-hand side of the chart, the community's perceived need to have access to this data decreases. This could be due to many factors including the difficulty in assuring the quality of such data as well as the lack of familiarity with tools to handle the data complexity. However, with complexity comes a richness of information that if properly tapped could be extraordinarily valuable. In astronomy, for example, the Sloan Digital Sky Survey created a very complex database of attributes of stars, galaxies, and

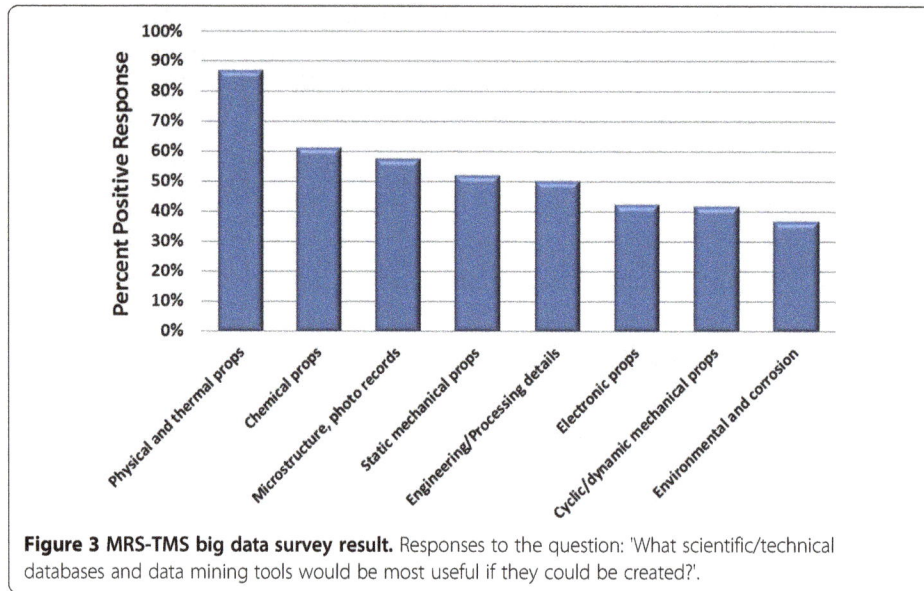

Figure 3 MRS-TMS big data survey result. Responses to the question: 'What scientific/technical databases and data mining tools would be most useful if they could be created?'.

quasars. The wealth of information and immense discovery potential led many in the research community to become expert users of SQL and for the survey to yield nearly 6,000 peer-reviewed publications.[a]

For those publications with wide technical scope, it will be difficult to provide a universal answer to 'what data should be archived?' In these cases, the decision for what data to archive may best be left to the judgment of the authors, peer reviewers, and editors. A particularly useful metric might be the cost/effort to produce the data. For example, the 'exquisite' experimental data associated with a high-energy diffraction microscopy experiment provide very unique, expensive, and rich datasets with great potential use to other researchers [25]. Clearly, based on these factors, the dataset should be archived. On the other hand, the results from a model run on commercial software that takes 5 min of desktop computation time may not be worthy of archiving as long as the input data, boundary conditions, and software version were well defined in the manuscript. Of course, one must account for the perishable nature of code, particularly old versions of commercial code. However, even the data from the common tensile test may be worthy of archiving as publications do not typically provide the entire curve; while the paper may report only yield strength, another researcher may be interested in work hardening behavior. Having the complete dataset in hand allows another researcher to explore alternative facets of the material's behavior. The basic elements of the criteria for determining the data required for archiving could include the following:

- Are the data central to the main scientific conclusions of the paper?
- Are the data likely to be usable by other scientists working in the field?
- Are the data described with sufficient pedigree and provenance that other scientists can reuse them in their proper context?
- Is the cost of reproducing the dataset substantially larger than the cost of archiving the fully curated dataset?
- Is the dataset reproducible at all, or does it stem from a unique event or experiment?

Data itself can come in a variety of 'processed' levels including 'raw', 'cleaned', and 'analyzed'. Such characterizations are subjective, though some disciplines have adopted quite rigorous definitions. Nonetheless, given the diversity of materials data, care will need to be taken in determining the appropriate amount of processing performed on a dataset to be archived. While raw or cleaned data is much preferred for its relative simplicity in reuse, it is probably much more important at this stage of our digital maturity that the metadata accompanying the dataset provide sufficient pedigree and provenance to make the data useful to others, including definition of the post-acquisition (experiment or computation) processing performed.

Another factor to consider in setting the guidelines for which data need to be archived is the expected annual and continuing storage capacity required. A very informal survey of 15 peer-reviewed journal article authors in NIST and Air Force Research Laboratory (AFRL) found that most articles in the survey had less than 2 GB of supporting data per paper. Currently, the time and resources required to upload (by authors) and download (by users) data files less than 2 GB are quite reasonable. However, those papers reporting on emerging characterization techniques such as 3D serial sectioning and high-energy diffraction microscopy were dependent on considerably larger datasets, approximately 500 GB per paper. Other disciplines have established data repositories to support their technical journals. Experience to date indicates that datasets of up to approximately 10 GB can be efficiently and cost effectively curated. Repositories such as the Dryad digital repository show that datasets of this magnitude can be indefinitely stored at a cost of $80 or less [13]. However, datasets approaching 500 GB will very likely require a different approach for storage and access. Thus, a data repository strategy needs to consider this range in distribution of datasets. An additional factor when considering long-term storage requirement is the high global rate of growth in materials science and engineering publications. Figure 4 shows the dramatic growth in the number of MSE journal articles published over the past two decades, indicating a commensurate amount of accompanying data.

Data repositories

Aside from crystallographic data repositories, there are at this time perhaps no dedicated materials data repositories that meet the required characteristics defined above. The materials science and engineering community does have numerous publicly accessible data repositories; however, the majority of these are associated with specific projects or research groups, and their persistence is therefore dependent on individual funding decisions. These repositories are primarily established to house and share the research data generated within a specific project or program. They generally do not follow uniform standards for data and metadata nor provide for data discoverability and citation. There are very few repositories established with the explicit objective of providing MSE with public repositories for accessible digital data. In short, publicly accessible, built-for-purpose repositories and the associated infrastructure for access, safe storage, and management still need to be developed and sustainably funded - this is the largest impediment to implementing viable data archiving policies. (See, for example, 'sustaining domain repositories for digital data: a white paper' [26]). In establishing their Joint Data Archiving Policy, journals in evolutionary biology did not prescribe specific repositories; instead, they allowed a mix of repositories to be used by authors as long as

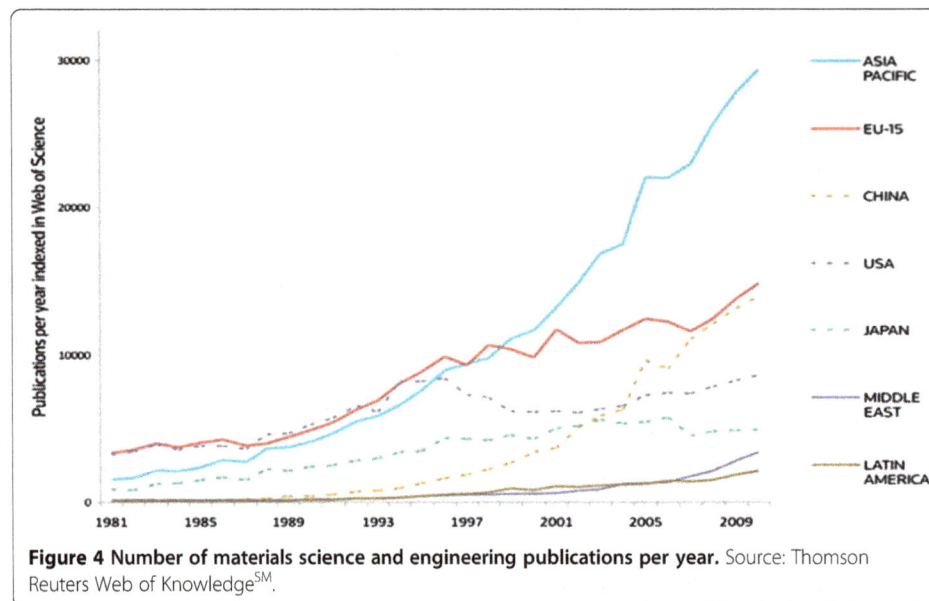

Figure 4 Number of materials science and engineering publications per year. Source: Thomson Reuters Web of Knowledge[SM].

they met established criteria. Such criteria may be as simple as requiring data cited to be permanently archived in data repositories that meet the following conditions:

- Publicly accessible throughout the world.
- Committed to archiving data sets indefinitely.
- Allow bi-directional linking between paper and dataset.
- Provide persistent digital identifier.

One tempting option might be to take advantage of the online storage capability several journals already offer for supplementary materials accompanying journal articles. However, as presently constructed, these are not amenable to best practices for dataset storage as they generally are not independently discoverable, searchable, separately citable, nor aggregated in one location. In fact, some publishers are reducing or eliminating supplementary file storage due to the haphazard structure and rules associated with their use. Further, new global government policies promoting open access to research works have the publishing industry in a state of flux with regard to their long-standing, subscription-based business model. Publishers have been reticent in taking on a data archiving responsibility given the economic uncertainties in the publishing marketplace.[b] Also, there is a possibility that some for-profit publishers could try to restrict access to digital data assets that are co-located with the journal.

As alluded to in the previous section, a fundamental consideration in repository design and/or selection is the level to which the repository will present structured versus unstructured data. Structured technical databases tend to be more useful to a technical community due to their uniformity, as evidenced by their data reuse rate.[c] A perfect construct would see the vast majority of materials data resident within structured repositories. A disciplined data structure provides enormous advantages to the researcher both in terms of data discoverability and confidence in its use. However, this structure must be enabled by the application of broader and deeper standards for data and metadata, standards that do not currently exist.

In all likelihood, as in biology, MSE publications will be dependent on a collection of repositories that are tailored to specific materials data. For example, NIST is building and demonstrating a data file repository for CALPHAD (Calculation of Phase Diagrams) and interatomic potentials [27]. These may be expandable and largely sufficient for thematic publications such as those devoted to thermodynamics and diffusion. However, repositories such as this will only fill a relatively small niche need in MSE. *Integrating Materials and Manufacturing Innovation* is piloting an effort to link articles with their supporting data using the NIST repository according to the criteria outlined above, an example can be found in an article by Shade et al. [28,29].

Finally, a business model for sustainably archiving materials data is required. Other technical fields, such as earth sciences, can at least partially rely on government-provided repositories for large and complex datasets. Without these types of repositories to build on, MSE will need to establish viable repository solutions. In response to funding-agency requirements for data management plans, some universities, Johns Hopkins for example, are beginning to provide centrally hosted data repositories, but these are not yet common [30]. Private fee-for-service repository services, such as LabArchives and Figshare, are also evolving to meet growing demand for accessible data storage [31,32][d]. Additionally, ASM International is working to create a prototype materials data repository through its close association with Granta Design. Termed the Computational Materials Data Network (CMDN), this is a promising option as the data repository will provide a structured database specifically for materials data; but the business model for CMDN has not yet been solidified [33]. A key open question remains how funding agencies will respond to the OSTP open research policy memo, and how they will fund activities making data open to the public.

Standards enabling data discoverability, exchange, and reuse

As noted in the previous section, standards for data and metadata provide the basis for a structured data archive, enabling the rapid discovery of data and assisting in determining the data's relevance and usefulness. At the most basic level, good data practice generally requires the generation, and acceptance, of a vocabulary defining the terms used to describe reported data. This assures the data user that they precisely understand the context of the data they are reviewing. From this level, other attributes, features, or requirements can be levied on a data management system including ontologies, schema, and formats [34].

Other fields have studied these issues as a community, and MSE is now reinvigorating a concerted effort to define its approach to setting data standards. Serious efforts to address standards for materials data, particularly structural materials data, were undertaken as long as 30 years ago [35]. In 1985, ASTM International established Committee E49 on Computerization of Material Property Data to develop standard guidelines and practices for materials databases [36]. ASTM International devoted quite substantial effort over a decade and issued data standards relevant to materials through the 1990s, specifically addressing key issues such as how to describe materials, how to record data, data quality indicators, harmonization of terminology, and guidelines for building and distributing databases. The standards have been since withdrawn, but those such as ASTM E1314-89, 'Practice for Structuring Terminological Records Relating to Computerized Test Reporting and Materials Design Formats', are clearly in need in today's environment - though in more web-enabled format [37]. ASTM International has been reviving its efforts to provide guidance on the digitization of materials test data by exploring the re-establishment of its computerization and networking of materials databases symposium series [38]. The European Union is studying the creation of standards for the exchange of engineering materials data through the European Committee for Standardization [39]. The target for these standards is structural materials with an early emphasis on aerospace applications. And the European Commission is funding a broader activity called the *Integrated Computational Materials Engineering expert group* (ICMEg) with the aim of developing the standards and protocols needed to support the digital exchange of materials data needed to conduct ICME [40]. Several recent papers have proposed standards for other types of materials data to include thermodynamic and image-based data [41,42]. There are also closed-loop approaches to materials data standardization that exist within commercial data management software packages, but these are not generally available to the public.

While the field of information technology is continuously evolving to provide solutions to more productively using unstructured data, at present there is no community-wide accepted practice for MSE data and metadata standards. Near-term solutions for governing the archiving of materials data will need to be relatively loose, flexible, and evolutionary with a drive toward more standardization. While publishers may not be able to directly provide data repository services, they are well positioned and willing to aid the community in establishment of data standards. In the pursuit of standardization across a technical field, Michael Whitlock, a primary champion of journal data archiving in the field of evolutionary biology, offered this quote from Voltaire based on his experience: 'the perfect is the enemy of the good'.[e] It is perhaps much more important at this stage of our digital maturity that MSE first implement data archiving with the best guidance available and work to build in standardization over time.

Data citation and attribution

Well-developed and uniform data citation standards are required to ensure that linkages between publications and datasets are enduring and that creators of digital datasets receive appropriate credit when their data are used by others. Standards for data citation practices and implementation provide the mechanism by which digital datasets can be reliably discovered and retrieved. Closely related to data citation, other challenges include the ability to reliably identify, locate, access, interpret, and verify the version, integrity, and provenance of digital datasets [43]. Any data archiving policy must concern itself not only with how publications should appropriately cite the datasets used but must also require attribution to authors of datasets outside the document.

Numerous organizations in the EU and USA have studied this issue and are continuing to refine technology solutions and best practices. For example, CODATA and the National Academy of Sciences released an in-depth international study and recommendations on citation of technical data [44]. Recently, these transnational initiatives have coalesced to produce a unified Joint Declaration of Data Citation Principles that is appropriate for any type of technical publication [45]. The eight principles define the purpose, function, and attributes of data citations and address the need for citations to be both understood by humans and processed by machines. With a slightly different perspective focused more on the mechanics of linking published articles with data repositories, DataCite and the International Association of Scientific, Technical and Medical

Publishers have issued a joint statement recommending best practices for citation of technical datasets in journals [46]. Two of the key recommendations include encouraging authors of research papers to deposit researcher validated data in trustworthy and reliable data archives and encouraging data archives to enable bi-directional linking between datasets and publications by using established and community-endorsed unique persistent identifiers such as database accession codes and digital object identifiers (DOIs).

An outstanding technical issue in data citation yet to be resolved concerns the granularity of the datasets used in a publication, both spatially and temporally. Spatial granularity refers to a subset of the dataset used in the research. Temporal granularity can refer to either the version of the dataset used or the temporal state of the dataset used if the dataset itself is dynamic.

Data quality

A key concern in linking datasets to publications is the provision of quality metrics; that is, can the data's ultimate reliability be assessed in a meaningful manner? Materials data can be provided as two basic types: experimental and computational; both types assume underlying models. In order for data and these associated models to be usable, their quality must be ascertained. In this context, it is useful to define the following for data and models:

- Pedigree - where did the information come from?
- Provenance - how was the information generated (protocols and equipment)? This metadata should be sufficient to reproduce the provided data.

In addition to these qualitative descriptors of the data, there are a number of meaningful quantitative measures of the data's quality. However, in general the following metrics are a strong basis for such an assessment:

- Verification - (applies to computational data only) how accurately does the computation solve the underlying equations of the model for the quantities of interest?
- Validation - how much agreement is there between realizations of a model in experiment and computational, or, rarely, analytic, results?
- Uncertainty - what is the quantitative level of confidence in our predictions?
- Sensitivity - how sensitive are results to changes in inputs or upon assumed boundary conditions?

Similar, and perhaps more difficult, problems pertain to simulation data. While such data may be perfectly precise in a numerical sense, simulations typically rely on many parameters, assumptions, and/or approximations. In principle, if the above are specified, and the quantitative metrics meet user requirements, the data can be used with a high level of confidence. A similar approach to defining data quality was recently proposed within the context of the Nanotechnology Knowledge Infrastructure Signature Initiative within the National Nanotechnology Initiative [47].

An often posed question in the research community with regard to data associated with peer-reviewed journal articles is that of peer review of the data itself. Indeed, it has been reported that approximately 50% of data being reviewed for submission to the American Mineralogist Crystal Structure Database contained errors [48]. The elements defined above represent the key criteria by which to judge the quality of the data. General pedigree and provenance information are typically conveyed in most research articles, though they may be provided in insufficient detail to reproduce the data. The remaining elements of validation, verification, uncertainty, and sensitivity are relatively loosely defined within materials science and engineering, and best practices have not generally been developed for each element, or, where developed, are not in widespread use. For further discussion on these topics, the reader is directed to [49] and [50].

Intellectual property and liability

There is quite a bit of complexity and even ambiguity with regard to the legal protections governing scientific data [51]. In general, scientific data are treated as facts and therefore not copyrightable under US law. However, the aggregation of the data into a single compilation or database may be copyrightable in the US. Additionally, and importantly, the codes, formats, metadata, data structures, or any 'added value' to the data could also be subject to copyright. Laws in other parts of the globe, particularly the European Union, add complexity to the situation. The EU's database directive, for example, protects the wholesale use of databases by other parties without permission.

There may be instances where the authors of a document may not want their data released immediately on publication of the supported manuscript. They may have very good, justifiable grounds to protect their data for some period following publication. One likely reason may be additional time required to file an invention disclosure related to the data. Another case may be that the authors are in the midst of writing another manuscript dependent on the same data. To account for these special cases, the associated publication should have allowance to grant the author an embargo period to protect the data for a short time after document publication. Typically, by granting an embargo, the author must post the supporting data to a repository prior to manuscript publication, but the data is not released to the public until the embargo period has expired. This is a standard practice in other technical disciplines, with limits of 12 months being typical and at the discretion of the editor.

Proprietary and export control restrictions may also affect the release of the metadata associated with the dataset and could warrant embargo or even permanent withholding of the entire metadata description. Take a researcher who has been provided a quantity of material by an industrial partner. The researcher may be free to report on a newly observed deformation phenomenon in the material with respect to its microstructure but may be restricted by the partner in providing proprietary details about how the material was processed. In this case, the metadata may not contain the full pedigree and provenance needed to reproduce the experimental results. Export control provides an analogous situation; the data may not be restricted, but the metadata needed to provide full pedigree and provenance may reveal export-controlled information [52]. Allowances for the withholding of metadata from publication must be in place, and these decisions to either accept the embargo or reject the dataset should be left to the reviewer and editor. It should be noted in publication policy that authors take full responsibility for review and release of proprietary and export-controlled information.

Given the discussion above regarding intellectual protection of data, policy regarding the requirements for licensure of data for reuse should be made clear. Of course, one must also consider where the data repository resides, so any policy may have somewhat limited scope. Creative Commons has developed a series of free copyright licenses for public use when sharing creative works [53]. For example, one option may be to require all deposited data to be covered by a CC-BY license, as defined by Creative Commons. A CC-BY grants free use of data by all parties, including for commercial use but does require attribution. However, other data repositories, such as the Dryad digital repository, have chosen to implement a Creative Commons Zero (CC0) license in order to remove any barriers to data reuse [13]. CC0 dedicates the work to the public domain and does not legally require attribution to the data source; instead Dryad relies on community norms for proper attribution of data. This option is particularly suitable in a case where a researcher uses data from hundreds or thousands of data sets in their work, making citation of all sources impractical. And, as noted at the beginning of this subsection, even the question of applicability of a copyright license to technical data is still open. Still, unanswered questions also linger regarding any liability issues with making data accessible. Again, consideration must be given to where the data reside (who is making it available) as to liability determination.

Archiving policy

A potential path forward is to establish a working group(s) comprised of members from the MSE community to craft a common data archiving policy. Such a policy should address the following:

1. A general definition of the data to be archived that is flexible enough to meet specific publication needs.
2. Criteria for suitable repositories.
3. Expectations or requirements to follow data or metadata standards.
4. Definition of standards for data citation and attribution.
5. Requirements and/or measures for data quality.
6. Clarity on intellectual property and liability issues.
7. Areas of opportunity for targeting pilot data archiving efforts (e.g., thermodynamic data).

Repositories

A complementary working group(s) from the MSE community should also be commissioned to develop a plan to provide supporting repositories for the MSE community. Some anticipated tasks and options include

1. Catalog and explore the suitability of and potential for existing materials repositories to host datasets associated with peer-reviewed journals (e.g., NIST CALPHAD database).
2. Explore the use of other established journal data repositories for their suitability for MSE data (e.g., www.datadryad.org).
3. Engage funding agencies for help in establishing specialized MSE data repositories.

4. Develop a time-phased strategy to provide well-structured materials repository architectures.

5. Consider business models that would sustain these repository services over the long term.

Conclusions

The era of open science is upon us, and the MSE community must generate a response that best suits the needs of not only the individual researcher but also the larger community including academia, industry, and government. It is becoming clearer with the advance of materials research that supporting data can no longer be kept invisible from a technical publication. This paper has outlined the key issues that will need to be considered as the community develops an approach to data archiving supporting publications. Charting the right course will take time, and much effort as it is quite complex. Fortunately, other technical disciplines have begun a path for us from which we can learn and capitalize. Some suggested community-based actions have been outlined that would help pave the way in setting a common approach to archiving of materials data.

Endnotes

[a]This is based on a query to the Astrophysics Data System, http://adsabs.harvard.edu/, for peer-reviewed papers mentioning either 'SSDS' or 'sloan' in the title or abstract of the paper. A query executed on 9 April 2014 resulted in 5,825 papers.

[b]CHW and JAW discussion with AAP, STM, AIP, ACS, Elsevier (2012).

[c]CHW discussion with A. Acharya, Google, Inc. (2012).

[d]'Certain commercial equipment, instruments, or materials (or suppliers, or software, ...) are identified in this paper to foster understanding. Such identification does not imply recommendation or endorsement by the US Government, nor does it imply that the materials or equipment identified are necessarily the best available for the purpose'.

[e]CHW discussion with M. Whitlock, U. British Columbia (2012).

Competing interests
The authors declare that they have no competing interests.

Authors' contributions
CHW structured the flow of the paper. CHW and JAW contributed a substantive portion of the manuscript, while RJH added valuable complementary perspectives from outside materials science and engineering throughout the subsections in the paper. All authors read and approved the final manuscript.

Acknowledgements
The authors wish to thank Clare Paul and Jeff Simmons for helpful discussions in preparing this manuscript. This article has undergone an impartial review process. Charles H. Ward excused himself from any tasks associated with the processing of this article. The review process was handled entirely by other editors and all decisions regarding this publication have been made by them.

Author details
[1]Air Force Research Laboratory, Materials and Manufacturing Directorate, Wright-Patterson AFB, OH 45433, USA. [2]National Institute of Standards and Technology, Gaithersburg, MD 20899, USA. [3]Space Telescope Science Institute, Baltimore, MD 21218, USA.

References
1. Materials Research Society and The Minerals, Metals and Materials Society (2013) MRS-TMS 'big data' survey. JOM 65:1073, doi:10.1007/s11837-013-0724-y

2. Committee on Integrated Computational Engineering (2008) Integrated computational materials engineering: a transformational discipline for improved competitiveness and national security. The National Academies Press, Washington, DC

3. National Institutes of Health (2003) Final NIH statement on sharing research data. http://grants.nih.gov/grants/guide/notice-files/NOT-OD-03-032.html. Accessed 13 May 2014

4. Department of Energy (2013) Policies on release of human genomic sequence data. http://web.ornl.gov/sci/techresources/Human_Genome/research/bermuda.shtml. Accessed 13 May 2014

5. National Science Foundation (2013) Grant proposal guide. http://www.nsf.gov/pubs/policydocs/pappguide/nsf13001/gpg_2.jsp#dmp. Accessed 13 May 2014

6. The White House (2011) The materials genome initiative. http://www.whitehouse.gov/mgi. Accessed 13 May 2014

7. Alliance for Permanent Access (2014) Opportunities for data exchange., http://www.alliancepermanentaccess.org/index.php/community/current-projects/ode/outputs/. Accessed 19 May 2014

8. The Royal Society (2012) Science as an open enterprise. https://royalsociety.org/~/media/policy/projects/sape/2012-06-20-saoe.pdf. Accessed 13 May 2014

9. RCUK Common Principles on Data Policy (2014) Research councils UK. http://www.rcuk.ac.uk/research/datapolicy/. Accessed 13 May 2014

10. Holdren JP (2013) Increasing access to the results of federally funded scientific research. http://www.whitehouse.gov/sites/default/files/microsites/ostp/ostp_public_access_memo_2013.pdf. Accessed 13 May 2014

11. The White House (2014) Project open data., http://project-open-data.github.io/. Accessed 13 May 2014

12. National Institutes of Health (2013) GenBank overview. http://www.ncbi.nlm.nih.gov/genbank/. Accessed 13 May 2014

13. Dryad digital repository (2014) Dryad. http://www.datadryad.org/. Accessed 13 May 2014

14. National Institute of Standards and Technology (2014) Thermodynamics Research Center. http://www.trc.nist.gov/. Accessed 13 May 2014

15. University of Leicester (2014) Peer REview for Publication & Accreditation of Research data in the Earth sciences [Online]. Available: http://www2.le.ac.uk/projects/preparde. Accessed 13 May 2014

16. The International Virtual Observatory Alliance (2014) Virtual observatory. http://ivoa.net/. Accessed 13 May 2014

17. Oxford Journals (2014) Database: The Journal of Biological Databases and Curation. http://www.oxfordjournals.org/our_journals/databa/about.html. Accessed 13 May 2014

18. Nature Publishing Group (2014) Scientific data. Macmillan. http://www.nature.com/scientificdata/. Accessed 13 May 2014

19. PLoS ONE editorial policies. PLoS One: http://www.plosone.org/static/editorial#sharing. Accessed 13 May 2014

20. Warren JA, Boisvert RF (2012) Building the materials innovation infrastructure: data and standards a Materials Genome Initiative Workshop. National Institute of Standards and Technology, Gaithersburg, MD, doi:10.6028/NIST.IR.7898

21. Whitlock MC, McPeek MA, Rausher MD, Rieseberg L, Moore AJ (2010) Data archiving. Am Nat 175:145–146, doi:10.1086/650340

22. Thomson Reuters (2014) Data citation index., http://thomsonreuters.com/data-citation-index/. Accessed 28 August 2014

23. (2012) Notes for authors. Acta Crystallogr C68:e3–e11, doi:10.1107/S0108270111047019

24. Koga N, Schick C, Vyazovkin S (2013) New procedures for articles reporting thermophysical properties. Thermochim Acta 555:iii, doi:10.1016/S0040-6031(13)00060-9

25. Miller M, Suter R, Lienert U, Beaudoin A, Fontes E, Almer E, Schuren J (2012) High-energy needs and capabilities to study multiscale phenomena in crystalline materials. Synchrotron Radiat News 25(6):18–26, doi:10.1080/08940886.2012.736834

26. Ember C, Hanisch R (2013) Sustaining domain repositories for digital data: a white paper. http://datacommunity.icpsr.umich.edu/sites/default/files/WhitePaper_ICPSR_SDRDD_121113.pdf. Accessed 13 May 2014

27. NIST Computational File Repository (2014) National Institute of Standards and Technology. http://nist.matdl.org/dspace/xmlui/handle/11115/52. Accessed 13 May 2014

28. Shade PA, Groeber MA, Schuren JC, Uchic MD (2013) Experimental measurement of surface strains and local lattice rotations combined with 3D microstructure reconstruction from deformed polycrystalline ensembles at the micro-scale. Integr Mater Manuf Innovation 2:5, doi:10.1186/2193-9772-2-5

29. Shade PA, Groeber MA, Schuren JC, Uchic MD (2013) 3D microstructure reconstruction of polycrystalline nickel micro-tension test. http://hdl.handle.net/11115/152 . Accessed 13 May 2014

30. Johns Hopkins University (2014) Research data management services at JHU. http://dmp.data.jhu.edu/. Accessed 15 May 2014

31. Labarchives (2014) LabArchives LLC. Carlsbad, CA, http://labarchives.com/. Accessed 13 May 2014

32. Figshare (2014) Figshare, London. http://figshare.com/. Accessed 13 May 2014

33. Computational Materials Data Network (2014) ASM International., http://www.cmdnetwork.org/content/cmdnetwork/. Accessed 13 May 2014

34. Cheung K, Hunter J, Drennan J (2009) MatSeek: an ontology-based federated search interface for materials scientists. IEEE Intell Syst 24:47–56, doi:10.1109/MIS.2009.13

35. Freiman S, Madsen L, Rumble J (2011) A perspective on materials databases. Am Ceram Soc Bull 90(2):28–32

36. Rumble J (1991) Standards for materials databases: ASTM Committee E49. In: Kaufman JG, Glatzman JS (eds) Computerization and networking of materials databases: Second Volume, ASTM STP 1106. American Society for Testing and Materials, Philadelphia, pp 73–83

37. ASTM International (1999) ASTM E1314-89(1999): Practice for structuring terminological records relating to computerized test reporting and materials design formats. ASTM International, West Conshohocken, PA

38. Rumble J (2014) E-Materials Data. ASTM International. Stand News. http://www.astm.org/standardization-news/perspective/ematerials-data-ma14.html. Accessed 15 May 2014

39. Austin T, Bullough C, Gagliardi D, Leal D, Loveday M (2013) Prenormative research into standard messaging formats for engineering materials data. Int J Dig Curation 8:5–13, doi:10.2218/ijdc.v8i1.245

40. Schmitz GJ, Prahl U (2014) ICMEg—the integrated computational materials engineering expert group—a new European coordination action. Integr Mater Manuf Innov 3:2, doi:10.1186/2193

41. Campbell CE, Kattner UR, Liu Z-K (2014) The development of phase-based property data using the CALPHAD method and infrastructure needs. Integr Mater Manuf Innov 3:12, doi:10.1186/2193-9772-3-12
42. Jackson MA, Groeber MA, Uchic MD, Rowenhorst DJ, De Graef M (2014) h5ebsd: an archival data format for electron back-scatter diffraction data sets. Integr Mater Manuf Innov 3:4, doi:10.1186/2193-9772-3-4
43. Uhlir PF (rapporteur) (2012) For attribution-developing data attribution and citation practices and standards. National Academies Press, Washington
44. Socha YM (2013) Out of sight, out of mind: the current state of practice, policy, and technology for the citation of data. Data Sci J 12f:13
45. Data Citation Synthesis Group (2014) Joint declaration of data citation principles. FORCE11. https://www.force11.org/datacitation. Accessed 14 May 2014
46. Data Cite and International Assocation of Scientific, Technical and Medical Publishers (2012) STM-DataCite joint statement. http://www.stm-assoc.org/2012_06_14_STM_DataCite_Joint_Statement.pdf. Accessed 14 May 2014
47. National Nanotechnology Initiative (2013) Nanotechnology knowledge infrastructure data readiness levels discussion draft. http://www.nano.gov/node/1015. Accessed 14 May 2014
48. Downs RT, Hall-Wallace M (2003) The American mineralogist crystal structure database. Am Mineral 88:247–250
49. Cowles B, Backman D, Dutton R (2012) Verification and validation of ICME methods and models for aerospace applications. Integr Mater Manuf Innov 1:2, doi:10.1186/2193-9772-1-2
50. Committee on Mathematical Foundations of Verification, Validation, and Uncertainty Quantification (2012) Assessing the reliability of complex models. The National Academies Press, Washington, DC
51. Madison M (2012) In: Uhlir PF (rapporteur) (ed) The future of scientific knowledge discovery in open networked environments. National Academies Press, Washington, pp 101–106
52. Ward CH (2013) Implications of integrated computational materials engineering with respect to export control. AFRL-RX-WP-TM-2013-0156. Defense Technical Information Center, Fort Belvoir, VA
53. Creative Commons (2014) Creative Commons. http://creativecommons.org/licenses/. Accessed 14 May 2014

4

Exploration of data science techniques to predict fatigue strength of steel from composition and processing parameters

Ankit Agrawal[1][*], Parijat D Deshpande[2], Ahmet Cecen[3], Gautham P Basavarsu[2], Alok N Choudhary[1] and Surya R Kalidindi[3,4]

*Correspondence:
ankitag@eecs.northwestern.edu
[1] Department of Electrical
Engineering and Computer Science,
Northwestern University, Evanston,
IL, USA
Full list of author information is
available at the end of the article

Abstract

This paper describes the use of data analytics tools for predicting the fatigue strength of steels. Several physics-based as well as data-driven approaches have been used to arrive at correlations between various properties of alloys and their compositions and manufacturing process parameters. Data-driven approaches are of significant interest to materials engineers especially in arriving at extreme value properties such as cyclic fatigue, where the current state-of-the-art physics based models have severe limitations. Unfortunately, there is limited amount of documented success in these efforts. In this paper, we explore the application of different data science techniques, including feature selection and predictive modeling, to the fatigue properties of steels, utilizing the data from the National Institute for Material Science (NIMS) public domain database, and present a systematic end-to-end framework for exploring materials informatics. Results demonstrate that several advanced data analytics techniques such as neural networks, decision trees, and multivariate polynomial regression can achieve significant improvement in the prediction accuracy over previous efforts, with R^2 values over 0.97. The results have successfully demonstrated the utility of such data mining tools for ranking the composition and process parameters in the order of their potential for predicting fatigue strength of steels, and actually develop predictive models for the same.

Keywords: Materials informatics; Data mining; Regression analysis; Processing-property linkages

Background

Causal relations are foundational to all advances in sciences and technology. In advancing materials science and engineering, the practitioners of the field have traditionally relied largely on observations made from cleverly designed controlled experiments and sophisticated physics-based models to establish the desired causal relations, e.g., process-structure-property (PSP) linkages. In recent vision-setting documents [1,2], experts in materials science and engineering have identified data science and analytics as offering a third set of distinct tools (i.e., experiments, models, and data analytics making up the three foundational components of an integrated approach) for establishing the desired causal relations. Data science and analytics is expected to positively impact the ongoing materials development efforts by maximizing the accuracy and reliability of the core knowledge mined from large ensembles of (often heterogeneous and/or incomplete) datasets, and providing clear guidance for investment of future effort in as yet unexplored "white" spaces with the highest potential for success/benefit. In fact, data analytics techniques have already been successfully making inroads in the quest for new material design and discovery. The considerable interest and progress in the recent years has resulted in developing this new field and terming it as "Materials Informatics" [3,4]. The Materials Genome Initiative [2] places a large emphasis on data-driven approaches. Progress in this direction is supplemented by the availability of large amounts of experimental and simulation data, enhanced computing tools and advances in data analytics, which is expected to augment rather than compete with existing analytics methods.

State-of-the-art data analytics

Over the last few decades, our ability to generate data has far exceeded our ability to make sense of it in practically all scientific domains, and materials science is no exception. This has led to the emergence of the fourth paradigm of science [5], which is data-driven science and discovery, and is based on developing predictive and discovery-based data mining approaches on big data in a comprehensive manner. Fourth paradigm compliments the three traditional scientific advancement models of mathematical modeling, experiments, and computer simulations. Indeed, the most advanced techniques in this field come from computer science, high-performance computing, machine learning and data mining algorithms, and via applications in business domain, climate science, bioinformatics, astronomy/cosmology, intrusion detection, network analysis and many others, where predictive data mining has been effectively used for decision making with significant gains in the outcomes relevant to that domain. For example, companies like Amazon [6,7], Netflix [8], Google [9], Walmart [10] and Target [11] use predictive modeling for recommendations, personalized news, cost reductions, predicting demand and supply chain management at a massive scale providing lifts in sales and satisfaction. Scientists use predictive mining on big data to discover new stars/galaxies, predict hurricane paths, or predict structure of new materials. The accurate prediction of the path of hurricane Sandy illustrates an example of how the use and analysis of much larger data sets and algorithms can significantly improve accuracy.

The impressive advances made in the last two decades in both materials characterization equipment and in the physics-based multiscale materials modeling tools have ushered the BIG DATA age of materials science and engineering. With the advent of big data came the recognition that advanced statistics and modern data analytics would have to play an important role in the future workflows for the development of new or improved materials.

Several case studies illustrating the potential benefits of this emerging new field of Materials Informatics (MI) have already been published in literature. Rajan et al. [12] applied principal component analysis (PCA) on a database consisting of 600 compounds of high temperature superconductors to identify patterns and factors which govern this important property. They observed that the dataset clusters according to the average valency, a criterion which has been reported in literature to be of utmost importance for superconducting property. They concluded that informatics techniques allow one to investigate complex multivariate information in an accelerated and yet physically meaningful manner. Suh and Rajan [13] applied informatics tools on a dataset consisting of AB2N4 spinel nitrides to find the statistical interdependency of factors that may influence chemistry-structure-property relationships. Using partial least squares (PLS), they developed a quantitative structure-activity relationship (QSAR) relating bulk modulus of AB2N4 spinels with a variety of control parameters. They observed a strong agreement between the properties predicted based on ab-initio calculations and the ones based strictly on a data-driven approach. Nowers et al. [14] investigated property-structure-processing relations during interpenetrating polymer network (IPN) formation in epoxy/acrylate systems using an informatics approach. They concluded that material informatics is a very efficient tool which can be utilized for additional materials development. Gadzuric et al. [15] applied informatics tools on molten salt database to predict enthalpy (δH_{form}) and Gibbs free energy of formation (δG_{form}) of lanthanide halides. The results of the analysis indicated a high level of confidence for the predictions. George et al. [16] applied similar approach on a dataset consisting of binary and ternary metal hydrides to investigate the interrelationships among material properties of hydrides. They developed a relationship between entropy of a hydride and its molar volume which was in close agreement with the theoretical predictions. Singh et al. [17] developed a neural network model in which the yield and tensile strength of the steel was estimated as a function of some 108 variables, including the chemical composition and an array of rolling parameters. Fujii et al. [18] applied neural network approach for prediction of fatigue crack growth rate of nickel base superalloys. They modeled the rate as a function of 51 variables and demonstrated the ability of such methods for investigation of new phenomena in cases where the information cannot be accessed experimentally. Hancheng et al. [19] developed an adaptive fuzzy neural network model to predict strength based on compositions and microstructure. Runway stiffness prediction and evaluation models have also been developed using techinques such as genetic programming [20] and artificial neural networks [21]. Wen et al. [22] applied support vector regression (SVR) approach for prediction of corrosion rate of steels under different seawater environments. They concluded that SVR is a promising and practical method for real-time corrosion tracking of steels. Rao et al. [23] applied SVR for prediction of grindability index of coal and concluded that SVR is a promising technique and needs smaller data set for training the model than artificial neural network (ANN) techniques. To the best of our knowledge, there is only one prior study [24] dealing with fatigue strength prediction using the NIMS database (same data that we use in this work; details of the data provided later). It applied PCA on the data and subsequently performed partial least square regression (PLSR) on the different clusters identified by PCA for making predictions. Large R^2 values ranging between 0.88 and 0.94 were obtained for the resulting clusters.

Motivation

The prior MI case studies cited above have established the unequivocal potential of this emerging discipline in accelerating discovery and design of new/improved materials. However, there still does not exist a standardized set of protocols for exploring this approach in a systematic manner on many potential applications, and thus, establishing the composition-processing-structure-property relationships still remains an arduous task. A report published by NRC [1] stated that materials design has not been able to keep pace with the product design and development cycle and that insertion of new materials has become more infrequent. This poses a threat to industries such as automotive and aerospace, in which the synergy between product design, materials, and manufacturing is a competitive advantage.

In this paper, we embark on establishing a systematic framework for exploring MI, and illustrate it by establishing highly reliable causal linkages between process variables in a class of steels, their chemical compositions, and their fatigue strengths. The approach described in this work comprises of four main steps: (i) Preprocessing for consistent description of data, which can include things like filling in missing data wherever possible, with the help of appropriate domain knowledge; ii) Feature selection for attribute ranking and/or identifying the best subset of attributes for establishing a given linkage; (iii) Predictive modeling using multiple statistical and advanced data-driven strategies for the establishment of the desired linkages, (iv) Critical evaluation of the different informatics approaches using appropriate metrics and evaluation setting to avoid model over-fitting.

Accurate prediction of fatigue strength of steels is of particular significance in materials science to several advanced technology applications because of the extremely high cost (and time) of fatigue testing and often debilitating consequences of fatigue failures. Fatigue strength is the most important and basic data required for design and failure analysis of mechanical components. It is reported that fatigue accounts for over 90% of all mechanical failures of structural components [25]. Hence, fatigue life prediction is of utmost importance to both the materials science and mechanical engineering communities. The unavailability of recorded research in using a large number of heat treatment process parameters, composition to predict extreme value properties such as fatigue strength has led us to work on this problem. The complex interaction between the various input variables have baffled the conventional attempts in pursuing this work and advanced data analytics techniques may lead the path. The aim of this study is thus to fill some of the gaps encountered in this work and serve as a preliminary guide for prospective researchers in this field. The scope of this paper includes application of a range of machine learning and data analytics methods applied for the problem of fatigue strength prediction of steels using composition and processing parameters. A conference version of this paper with preliminary results appeared in Proceedings of the 2nd World Congress on Integrated Computational Materials Engineering (ICME 2013) [26].

Data

Fatigue Dataset for Steel from National Institute of Material Science (NIMS) MatNavi [27] was used in this work, which is one of the largest databases in the world with details on composition, mill product (upstream) features and subsequent processing (heat treatment) parameters. The database comprises carbon and low-alloy steels, carburizing steels and spring steels. Fatigue life data, which pertain to rotating bending fatigue tests at room temperature conditions, was the target property for which we aimed to construct predictive models in the current study. The features in the dataset can be categorized into the following:

- Chemical composition - %C, %Si, %Mn, %P, %S, %Ni, %Cr, %Cu, %Mo (all in wt. %)
- Upstream processing details - ingot size, reduction ratio, non-metallic inclusions

- Heat treatment conditions - temperature, time and other process conditions for normalizing, through-hardening, carburizing-quenching and tempering processes
- Mechanical properties - YS, UTS, %EL, %RA, hardness, Charpy impact value (J/cm2), fatigue strength.

The data used in this work has 437 instances/rows, 25 features/columns (composition and processing parameters), and 1 target property (fatigue strength). The 437 data instances include 371 carbon and low alloy steels, 48 carburizing steels, and 18 spring steels. This data pertains to various heats of each grade of steel and different processing conditions. The details of the 25 features and given in Table 1.

Methods

The overall proposed approach is illustrated in Figure 1. The raw data is preprocessed for consistency using domain knowledge. Ranking-based feature selection methods are also used to get an idea of the relative predictive potential of the attributes. Different regression-based predictive modeling methods are then used on the preprocessed and/or transformed data to construct models to predict the fatigue strength, given the composition and processing parameters. All constructed models are evaluated using Leave-One-Out Cross Validation with respect to various metrics for prediction accuracy. Below we present the details of each of the 4 stages.

Table 1 NIMS data features

Abbreviation	Details
C	% Carbon
Si	% Silicon
Mn	% Manganese
P	% Phosphorus
S	% Sulphur
Ni	% Nickel
Cr	% Chromium
Cu	% Copper
Mo	% Molybdenum
NT	Normalizing Temperature
THT	Through Hardening Temperature
THt	Through Hardening Time
THQCr	Cooling Rate for Through Hardening
CT	Carburization Temperature
Ct	Carburization Time
DT	Diffusion Temperature
Dt	Diffusion time
QmT	Quenching Media Temperature (for Carburization)
TT	Tempering Temperature
Tt	Tempering Time
TCr	Cooling Rate for Tempering
RedRatio	Reduction Ratio (Ingot to Bar)
dA	Area Proportion of Inclusions Deformed by Plastic Work
dB	Area Proportion of Inclusions Occurring in Discontinuous Array
dC	Area Proportion of Isolated Inclusions
Fatigue	**Rotating Bending Fatigue Strength (10^7 Cycles)**

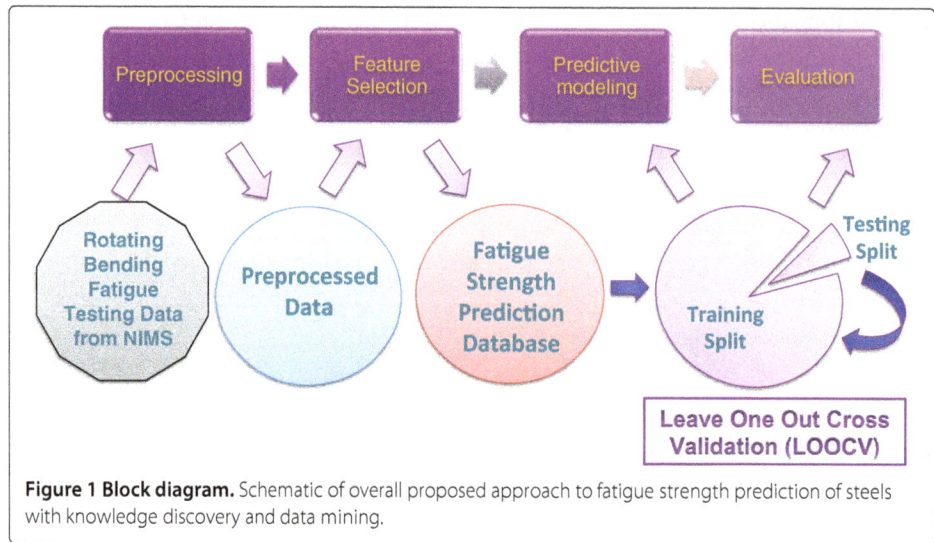

Figure 1 Block diagram. Schematic of overall proposed approach to fatigue strength prediction of steels with knowledge discovery and data mining.

Preprocessing

Understanding and cleaning the data for proper normalization is one of the most important steps for effective data mining. Appropriate preprocessing, therefore, becomes extremely crucial in any kind of predictive modeling, including that of fatigue strength. The dataset used in this study consists of multiple grades of steel and in some records, some of the heat treatment processing steps did not exist. In particular, different specimens are subjected to different heat treatment conditions. For example, some are normalized and tempered, some are through hardened and tempered, and others are carburized and tempered. There could be cases where normalization is done prior to carburization and tempering. In order to bring in a structure to the database, we have included all the key processes in the data-normalization, through hardening, carburization, quenching and tempering. For the cases where the actual process does not take place, we set the appropriate duration/time variable to zero with corresponding temperature as the austenization temperature or the average of rest of the data where the process exists. Setting the time to zero would essentially mean that no material transformation occurs. An artifact of our resulting data is that we are treating temperature and time as independent variables whereas they actually make sense only when seen together.

Feature selection

Information gain

This is an entropy-based metric that evaluates each attribute independently in terms of its worth by measuring the information gain with respect to the target variable:

$$IG(Class, Attrib) = H(Class) - H(Class|Attrib) \tag{1}$$

where $H(.)$ denotes the information entropy. The ranking generated by this method can be useful to get insights about the relative predictive potential of the input features.

SVD-PCA

Singular value decomposition is a matrix factorization defined as:

$$D = U \times S \times V \tag{2}$$

where, D is the data matrix such that every observation is represented by a row and each column is an explanatory variable, U is the matrix of left singular vectors, V is the matrix of right singular vectors and S is the diagonal matrix of singular values. In this case, $A = U \times S$ is a transformation of D where the data is represented by a new set of explanatory variables such that each variable is a known linear combination of the original explanatory parameters. The dimensions of A are also referred to as the Principal Components (PC) of the data.

Predictive modeling

We experimented with 12 predictive modeling techniques in this research study, which include the following:

Linear regression

Linear regression probably the oldest and most widely used predictive model, which commonly represents a regression that is linear in the unknown parameters used in the fit. The most common form of linear regression is least squares fitting [28]. Least squares fitting of lines and polynomials are both forms of linear regression.

Pace regression

It evaluates the effect of each feature and uses a clustering analysis to improve the statistical basis for estimating their contribution to overall regression. It can be shown that pace regression is optimal when the number of coefficients tends to infinity. We use a version of Pace Regression described in [29,30].

Regression post non-linear transformation of select input variables

A non-linear transformation of certain input variables can be done and the resulting data-set used for linear regression. In this study, the temperature variation effects on the diffusion equation are modelled according to the Arrhenius' empirical equation as $exp(-1/T)$ where T is measured in Kelvin.

Robust fit regression

The robust regression method [31] attempts to mitigate the shortcomings which are likely to affect ordinary linear regression due to the presence of outliers in the data or non-normal measurement errors.

Multivariate polynomial regression

Ordinary least squares (OLS) regression is governed by the equation:

$$\beta = (X'X)^{-1}X'Y \tag{3}$$

where β is the vector of regression coefficients, X is the design matrix and Y is the vector of responses at each data point. Multivariate Polynomial Regression (MPR) is a specialized instance of multivariate OLS regression that assumes that the relationship between regressors and the response variable can be explained with a standard polynomial. Standard polynomial here refers to a polynomial function that contains every polynomial term implied by a multinomial expansion of the regressors with a given degree (sometimes also referred to as a polynomial basis function). Polynomials of various degrees and number of variables are interrogated systematically to find the most suitable fit. There is a finite number of possible standard polynomials that can be interrogated due to the degree of freedom imposed by a particular dataset; the number of terms in the polynomial (consequently the number of coefficients) cannot exceed the number of data points.

Instance-based

This is a lazy predictive modeling technique which implements the K-nearest-neighbour (kNN) modeling. It uses normalized Euclidean distance to find the training instance closest to the given test instance, and predicts the same class as this training instance [32]. If multiple instances have the same (smallest) distance to the test instance, the first one found is used. It eliminates the need for building models and supports adding new instances to the training database dynamically. However, the zero training time comes at the expense of a large amount of time for testing since each test instance needs to be compared with all the data instances in the training data.

KStar

KStar [33] is another lazy instance-based modeling technique, i.e., the class of a test instance is based upon the class of those training instances similar to it, as determined by some similarity function. It differs from other instance-based learners in that it uses an entropy-based distance function. The underlying technique used of summing probabilities over all possible paths is believed to contribute to its good overall performance over certain rule-based and instance-based methods. It also allows an integration of both symbolic and real valued attributes.

Decision table

Decision table is a rule-based modeling technique that typically constructs rules involving different combinations of attributes, which are selected using an attribute selection search method. It thus represents one of the simplest and most rudimentary ways of representing the output from a machine learning algorithm, showing a decision based on the values of a number of attributes of an instance. The number and specific types of attributes can vary to suit the needs of the task. Simple decision table majority classifier [34] has been shown to sometimes outperform state-of-the-art classifiers. Decision tables are easy for humans to understand, especially if the number of rules are not very large.

Support vector machines

SVMs are based on the Structural Risk Minimization (SRM) principle from statistical learning theory. A detailed description of SVMs and SRM is available in [35]. In their basic form, SVMs attempt to perform classification by constructing hyperplanes in a multidimensional space that separate the cases of different class labels. It supports both classification and regression tasks and can handle multiple continuous and nominal variables. Different types of kernels can be used in SVM models, including linear, polynomial, radial basis function (RBF), and sigmoid. Of these, the RBF kernel is the most recommended and popularly used, since it has finite response across the entire range of the real x-axis.

Artificial neural networks

ANNs are networks of interconnected artificial neurons, and are commonly used for non-linear statistical data modeling to model complex relationships between inputs and outputs. The network includes a hidden layer of multiple artificial neurons connected to the inputs and outputs with different edge weights. The internal edge weights are 'learnt' during the training process using techniques like back propagation. Several good descriptions of neural networks are available [36,37].

Reduced error pruning trees

A Reduced Error Pruning Tree (REPTree) [38] is an implementation of a fast decision tree learner. A decision tree consists of internal nodes denoting the different attributes and the branches denoting the possible values of the attributes, while the leaf nodes indicate the final predicted value of the target variable. REPTree builds a decision/regression tree using information gain/variance and prunes it using reduced-error pruning. In general, a decision tree construction begins at the top of the tree (root node) with all of the data. At each node, splits are made according to the information gain criterion, which splits the data into corresponding branches. Computation on remaining nodes continues in the same manner until one of the stopping criterions is met, which include maximum tree depth, minimum number of instances in a leaf node, minimum variance in a node.

M5 model trees

M5 Model Trees [39] are a reconstruction of Quinlan's M5 algorithm [40] for inducing trees of regression models, which combines a conventional decision tree with the option of linear regression functions at the nodes. It tries to partition the training data using a decision tree induction algorithm by trying to minimize the intra-subset variation in the class values down each branch, followed by back pruning and smoothing, which substantially increases prediction performance. It also uses the techniques used in CART [41] to effectively deal with enumerated attributes and missing values.

Evaluation

Traditional regression-based methods such as linear regression are typically evaluated by building the model (a linear equation in the case of linear regression) on the entire available data, and computing prediction errors on the same data. Although this approach works well in general for simple regression methods, it is nonetheless susceptible to overfitting, and thus can give over-optimistic accuracy numbers. In particular, a data-driven model can, in principle learn every single instance of the dataset and thus result in 100% accuracy on the same data, but will most likely not be able to work well on unseen data. For this reason, advanced data-driven techniques that usually result in black-box models need to be evaluated on data that the model has not seen while training. A simple way to do this is to build the model only on random half of the data, and use the remaining half for evaluation. This is called the train-test split setting for model evaluation. Further, the training and testing halves can then also be swapped for another round of evaluation and the results combined to get predictions for all the instances in the dataset. This setting is called 2-fold cross validation, as the dataset is split into 2 parts. It can further be generalized to k-fold cross validation, where the dataset is randomly split into k parts. $k-1$ parts are used to build the model and the remaining 1 part is used for testing. This process is repeated k times with different test splits, and the results combined to get preductions for the all the instances in the dataset using a model that did not see them while training. Cross validation is a standard evaluation setting to eliminate any chances of over-fitting.

Of course, k-fold cross validation necessitates builing k models, which may take a long time on large datasets.

Leave-one-out cross validation

We use leave-one-out cross validation (LOOCV) to evaluate and compare the prediction accuracy of the models. LOOCV is commonly used for this purpose particularly when the dataset is not very large. It is a special case of the more generic k-fold cross validation, with $k = N$, the number of instances in the dataset. The basic idea here is to estimate the accuracy of the predictive model on unseen input data it may encounter in the future, by withholding part of the data for training the model, and then testing the resulting model on the withheld data. In LOOCV, to predict the target attribute for each data instance, a separate predictive model is built using the remaining $N-1$ data instances. The resulting N predictions can then be compared with the N actual values to calculate various quantitative metrics for accuracy. In this way, each of the N instances is tested using a model that did not see it while training, thereby maximally utilizing the available data for model building, and at the same time eliminating the chances of over-fitting of the models.

Evaluation metrics

Quantitative assessments of the degree to how close the models could predict the actual outputs are used to provide an evaluation of the models' predictive performances. A multi-criteria assessment with various goodness-of-fit statistics was performed using all the data vectors to test the accuracy of the trained models. The criteria that are employed for evaluation of models' predictive performances are the coefficient of correlation (R), explained variance (R^2), Mean Absolute Error (MAE), and Root Mean Squared Error ($RMSE$), Standard Deviation of Error (SDE) between the actual and predicted values. The last three metrics were further normalized by the actual fatigue strength values to express them as error fractions. The definitions of these evaluation criteria are as follows:

$$R = \frac{\sum_{i=1}^{N}(y_i - \bar{y})(\hat{y}_i - \bar{\hat{y}})}{\sqrt{\sum_{i=1}^{N}(y_i - \bar{y})^2 \sum_{i=1}^{N}(\hat{y}_i - \bar{\hat{y}})^2}} \tag{4}$$

$$MAE = \bar{e} = \frac{1}{N}\sum_{N}|y - \hat{y}| \tag{5}$$

$$RMSE = \sqrt{\frac{1}{N}\sum_{N}(y - \hat{y})^2} \tag{6}$$

$$SDE = \sqrt{\frac{1}{N}\sum_{N}(|y - \hat{y}| - \bar{e})^2} \tag{7}$$

$$MAE_f = \bar{e_f} = \frac{1}{N}\sum_{N}\left|\frac{y - \hat{y}}{y}\right| \tag{8}$$

$$RMSE_f = \sqrt{\frac{1}{N}\sum_{N}\left(\frac{y - \hat{y}}{y}\right)^2} \tag{9}$$

$$SDE_f = \sqrt{\frac{1}{N}\sum_{N}\left(\left|\frac{y - \hat{y}}{y}\right| - \bar{e_f}\right)^2} \tag{10}$$

where y denotes the actual fatigue strength values (MPa), $\hat{y}$ denotes the predicted fatigue strength values (MPa), and N is the number of instances in the dataset.

The square of the coefficient of correlation, R^2, represents the variance explained by the model (higher the better), and is considered one of the most important metrics for evaluating the accuracy of regressive prediction models. Another useful metric is the fractional mean absolute error, MAE_f, which represents the error rate (lower the better).

Results and Discussion

We used the following statistical and data mining software for conducting the analysis reported in this paper: R [42], MATLAB [43], WEKA [44]. Default parameters were used unless stated otherwise.

The entire available data-set was assessed for visible clustering by employing K-means clustering technique. The cluster plot demonstrates inherent clustering in the available data, which agrees with the a priori knowledge of the dataset. The distinct clustering in the available data represents 3 clusters according to the grade of steels as depicted in Figure 2. These clusters however do not offer sufficient data-points to create individual meta-models for each cluster and hence, for all methods used, the entire data-set is used to develop predictive models.

The information gain metric was calculated for each of the 25 input attributes. For this purpose, the numerical fatigue strength was discretized into 5 equal-width bins as this only works for categorical target attributes. The relative predictive power of the 25 input attributes is shown in Figure 3. All the attributes were retained for building various predictive models as all of them were found to have significant predictive potential. We also looked at the correlation values of the 25 input features with the fatigue strength, as shown in Figure 4. Interestingly, TT (tempering temperature) shows up as the most important attribute for predicting fatigue strength in Figure 3. This is because the dataset

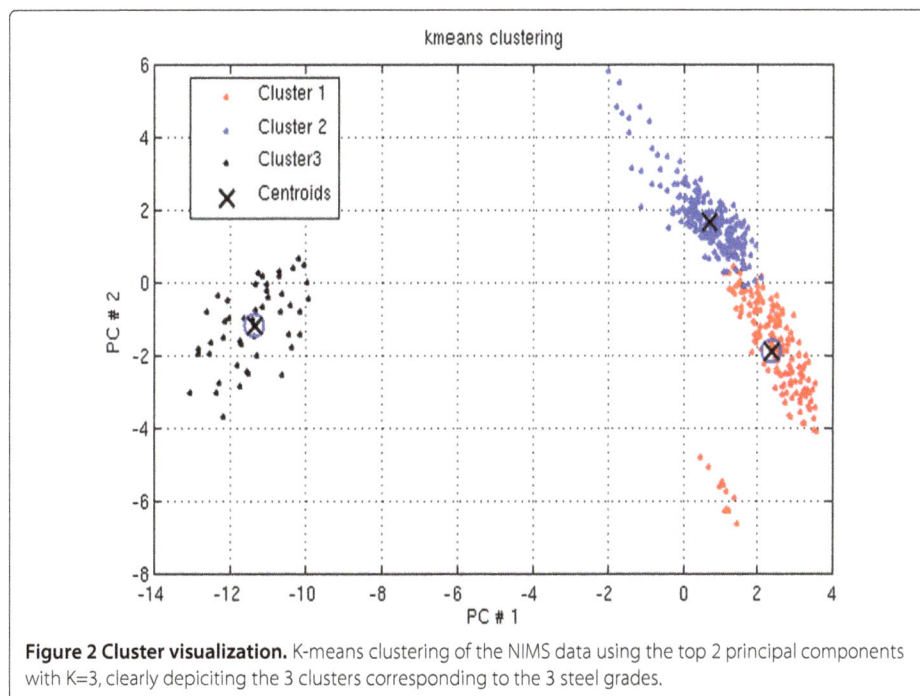

Figure 2 Cluster visualization. K-means clustering of the NIMS data using the top 2 principal components with K=3, clearly depicting the 3 clusters corresponding to the 3 steel grades.

Relative Attribute Importance

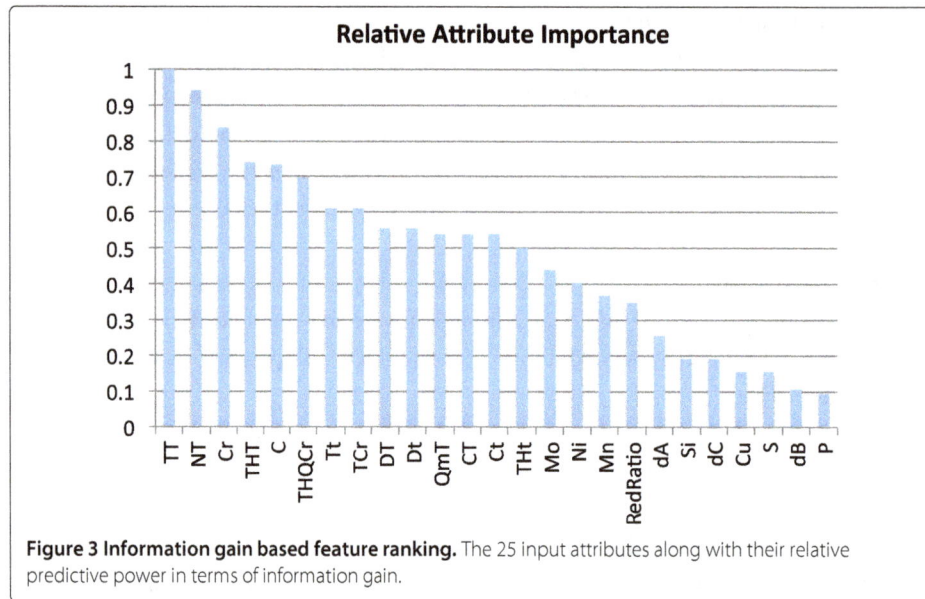

Figure 3 Information gain based feature ranking. The 25 input attributes along with their relative predictive power in terms of information gain.

consists of multiple grades of steel, each with a narrow yet significantly different range of TT. For example, TT for Through-hardened-tempered (without carburization) is around 400°C and that with carburization is around 200°C. These two situations will lead to a large difference in the fatigue strength. Thus, there is no surprise that the influence of TT seems high. However, the truth is that having a carburization step is what makes the key difference in the fatigue strength. Nevertheless, tempering will have a significant effect and this is reflected by the influence of tempering time in Figure 4. Figure 4 also identifies other variables such as carburizing temperature or through hardening temperature as important influencing factors. These are in line with expected results.

As mentioned before, we use Leave-One-Out Cross Validation (LOOCV) for model evaluation. Figure 5 and Table 2 present the LOOCV prediction accuracy of the 12 modeling techniques used in this work, in terms of the metrics discussed earlier. Clearly, many

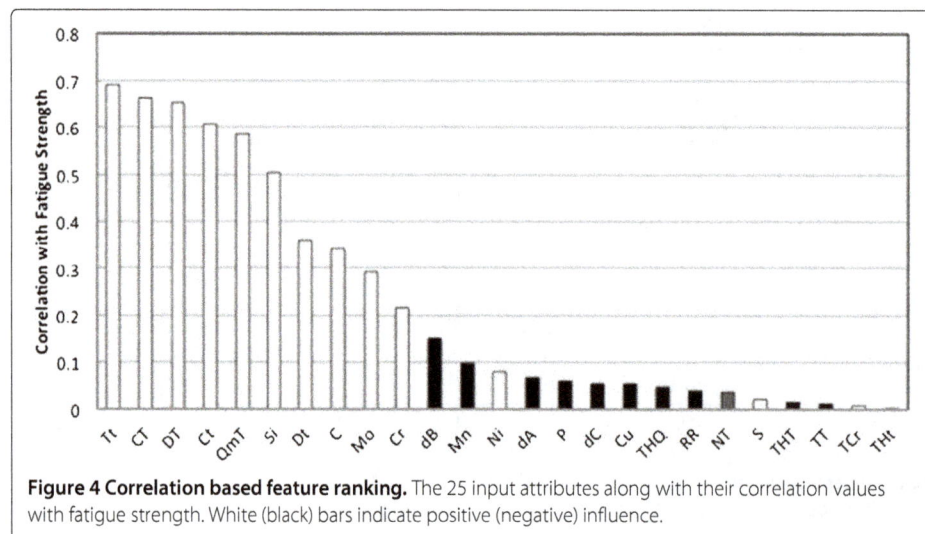

Figure 4 Correlation based feature ranking. The 25 input attributes along with their correlation values with fatigue strength. White (black) bars indicate positive (negative) influence.

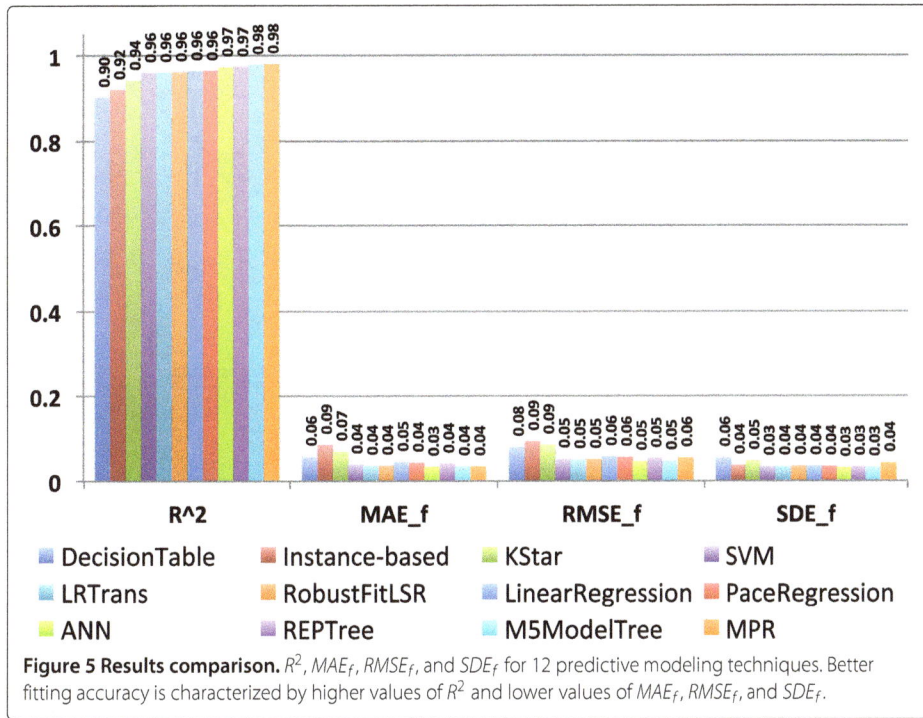

Figure 5 Results comparison. R^2, MAE_f, $RMSE_f$, and SDE_f for 12 predictive modeling techniques. Better fitting accuracy is characterized by higher values of R^2 and lower values of MAE_f, $RMSE_f$, and SDE_f.

of the employed data analytics techniques are able to achieve a high predictive accuracy, with R^2 values ~0.98, and error rate <4%. This is extremely encouraging since it significantly outperforms the only prior study on fatigue strength prediction [24], which reported R^2 values of <0.94. It is well known in the field of predictive data analytics that it becomes progressively more and more challenging to increase the accuracy of prediction beyond a certain point. To put it in context of this study, an increase in R^2 from 0.94 to 0.98 should not be viewed as simply an improvement of 0.04 or 4%. Rather, it should be seen with respect to the available scope for improvement of 0.06 (= 1.00 - 0.94). Thus, a more reasonable evaluation of the improvement accomplished by the current study over prior work would be about 66% (0.04/0.06), which is very significant.

Figure 6 presents the scatter plots for the 12 techniques. As can be seen from these plots, the three grades of steels are well separated in most of the techniques, and different

Table 2 Results comparison

Method	R	R^2	MAE	RMSE	SDE	MAE_f	$RMSE_f$	SDE_f
DecisionTable	0.9494	0.9014	34.8762	58.5932	47.1371	0.0584	0.0806	0.0557
IBk	0.9589	0.9195	46.0320	53.2749	26.8499	0.0859	0.0940	0.0382
KStar	0.9702	0.9413	36.9986	45.3779	26.3029	0.0706	0.0857	0.0487
SVM	0.9795	0.9594	24.2820	37.6250	28.7736	0.0400	0.0530	0.0349
LRTrans	0.9796	0.9596	22.3336	37.4748	30.1272	0.0370	0.0514	0.0357
RobustFitLSR	0.9804	0.9612	22.2152	37.2188	29.8960	0.0369	0.0520	0.0366
LinearRegression	0.9815	0.9633	25.6006	35.7168	24.9345	0.0456	0.0581	0.0360
PaceRegression	0.9816	0.9635	25.0302	35.5733	25.3065	0.0439	0.0565	0.0356
ANN	0.9861	0.9724	19.7778	31.0545	23.9695	0.0343	0.0470	0.0322
REPTree	0.9862	0.9726	22.5671	30.9401	21.1907	0.0414	0.0542	0.0349
M5ModelTree	0.9890	0.9781	19.3760	27.6065	19.6870	0.0353	0.0484	0.0332
MPR	0.9900	0.9801	18.5529	26.4378	18.8563	0.0350	0.0556	0.0432

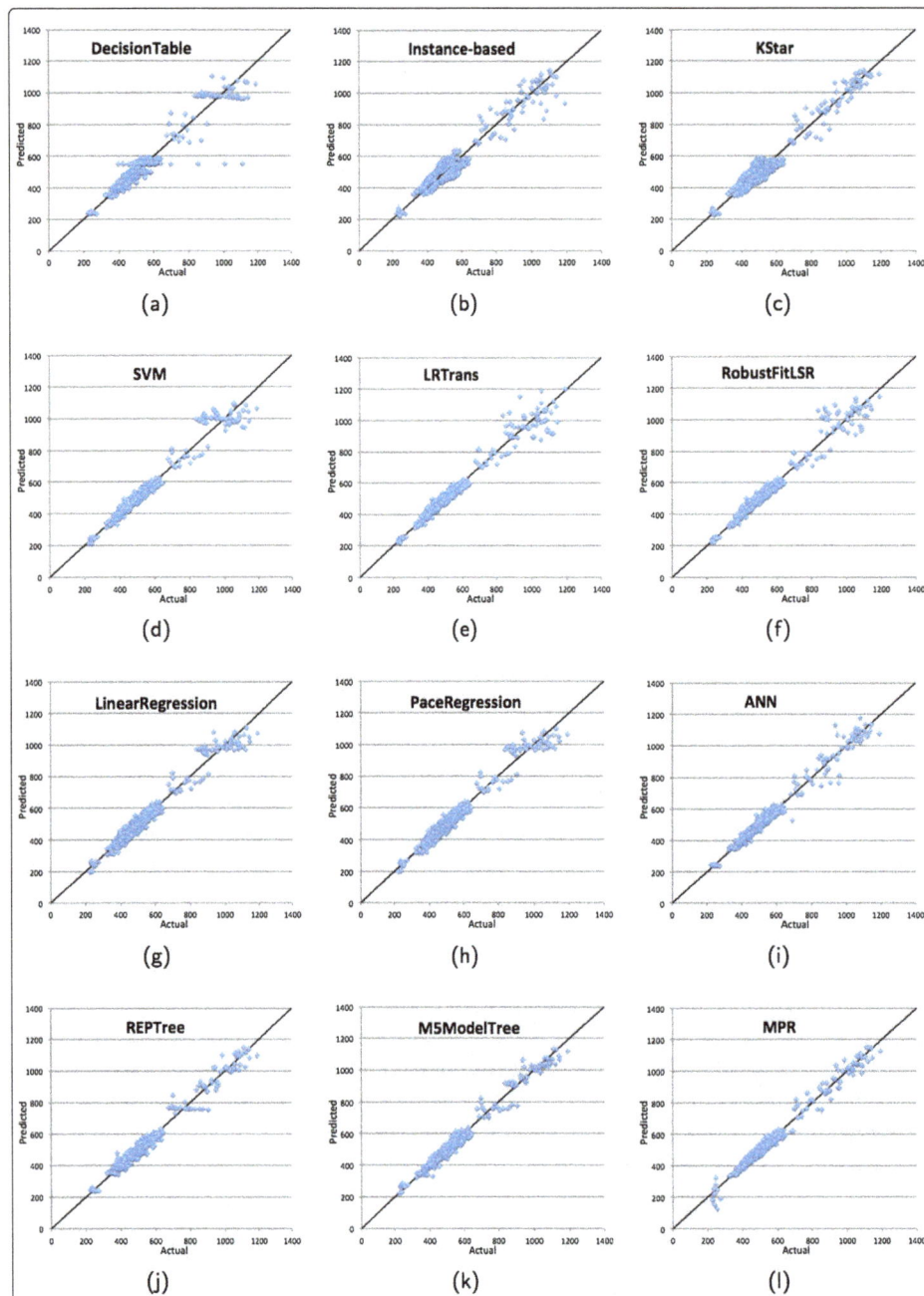

Figure 6 Scatter plots. Scatter plots for the 12 modeling techniques. X-axis and Y-axis denote the actual and predicted fatigue strength (in MPa) respectively. **a)** Decision Table; **b)** Instance-based; **c)** KStar; **d)** Support Vector Machines; **e)** Regression with Transformed Terms; **f)** RobustFit Regression; **g)** Linear Regression; **h)** Pace Regression; **i)** Artificial Neural Networks; **j)** Reduced Error Pruning Trees; **k)** M5 Model Trees; **l)** Multivariate Polynomial Regression.

techniques tend to perform better for different grades. Figure 7 shows the histograms of the error fractions for each of the techniques, to visualize the spread in the prediction errors. As expected, the spread in the error reduces as R^2 values improve. However, a point to be noted is that even though R^2 value is high, there are regions of data clusters where data fit is not sufficiently high and this is reflected in the nature of distribution of

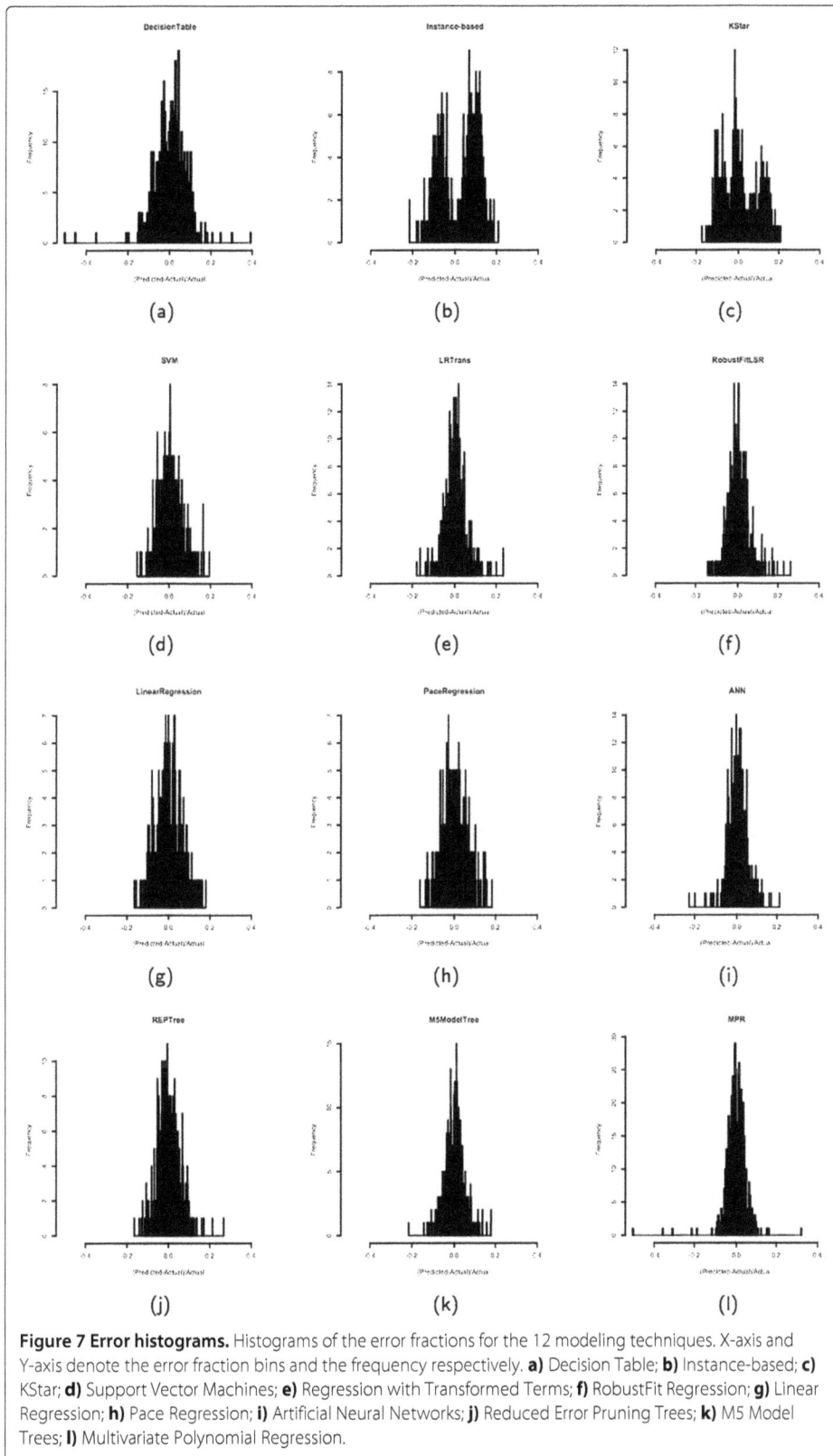

Figure 7 Error histograms. Histograms of the error fractions for the 12 modeling techniques. X-axis and Y-axis denote the error fraction bins and the frequency respectively. **a)** Decision Table; **b)** Instance-based; **c)** KStar; **d)** Support Vector Machines; **e)** Regression with Transformed Terms; **f)** RobustFit Regression; **g)** Linear Regression; **h)** Pace Regression; **i)** Artificial Neural Networks; **j)** Reduced Error Pruning Trees; **k)** M5 Model Trees; **l)** Multivariate Polynomial Regression.

errors. Thus, the methods that result in bimodal distribution of errors or the ones with significant peaks in higher error regions are not so good even though their reported R^2 may be reasonable.

The general opinion in data mining community about predictive modeling is that it is more helpful to know about a set of well performing techniques for a given problem rather than identifying a single winner. We have thus examined 12 different techniques for predictive modeling of fatigue strength, and it is shown that a number of different approaches produce highly reliable linkages. In particular, neural networks, decision trees, and multivariate polynomial regression were found to achieve a high R^2 value of greater than 0.97, which is significantly better than what has been previously reported in the literature. This is also shown by narrow distribution of errors. It is very encouraging to see that despite the limited amount of data available in this dataset, the data-driven analytics models were able to achieve a reasonably high degree of accuracy.

Although the main contribution of this paper is to present an end-to-end framework for exploring predictive materials informatics, and its application on NIMS data is a specific example of the application of the framework, it is nonetheless important for completeness to discuss some of the limitations of the proposed framework's specific application on the NIMS dataset. Since the data used in this study is very small compared to the typical amounts of data used in data mining studies in other domains, we believe that the obtained high accuracy is but an encouragement to use more data (possibly combine data from heterogenous sources) to further validate the results and/or making the model more robust. One possibility would be to add structure information to the data, which may ease the application of the developed models to actionable materials design, as structure information is what is primary responsible for the resulting properties. Another limitation of the NIMS data used in this study is the significantly different number of data instances corresponding to the different types of steels. Hence the predictive models, which are developed over the entire data may not be highly accurate for all steel types, which is also evident from the scatter plots. Possible approaches to deal with this imbalanced data distribution are discussed in the next section.

Conclusions

Materials Informatics, steeped in modern data analytics and advanced statistics, is fast emerging as a key enabler for accelerated and cost-effective development of new and improved materials targeted for advanced technologies. One of the core challenges addressed by this nascent field is the successful mining of highly reliable, quantitative, linkages capturing the salient connections between chemical compositions, processing history, and the final properties of the produced material. These linkages can provide valuable guidance to future effort investment with tremendous potential for cost-savings.

In this paper, we have tried to critically explore the viability of extracting such linkages from open access databases. As a specific example, we have focused on extracting reliable linkages between chemical compositions, processing history, and fatigue strength of a class of steels using data available from the open access materials database hosted by Japan's National Institute for Materials Science (NIMS). In this study, a range of advanced data analytics techniques, typically involving a combination of feature selection and regression methods, have been successfully employed and critically evaluated for the problem of fatigue strength prediction of different grades of steels.

There are several directions of future work that can stem from the present research. From the data analytics point of view, ensemble predictive modeling can be used to combine the results from multiple predictive models using same and/or different techniques built on different random subsets of the training data. Apart from this, since the scatter plots showed that different techniques can work well for different steel types, we can also try hierarchical predictive modeling, where we first try to classify the input test instance into one of the three grades of steel, and subsequently use the appropriate model(s) for that grade to predict the fatigue strength. From the materials science point of view, it would be good to explore the use of additional input features that may be easily measurable like some mechanical properties. Methods for using grouped variables representing each processing step could be of significant utility as well. It would also be extremely valuable to add structure information to the data, which may be able to give more actionable insights for materials design, as structure is very closely linked to property. Finally, the analytics framework developed and used in this paper can be used for building prediction models for other desired target properties, such as % Elongation, the data for which is also available in the NIMS dataset.

Availability of supporting data

The raw data used in this study was obtained from the publicly available NIMS MatNavi dataset, and was preprocessed as described in the paper. The preprocessed data is available as Additional file 1 of this paper.

Additional file

Additional file 1: Preprocessed NIMS MatNavi Data used in this study.

Competing interests
The authors declare that they have no competing interests.

Authors' contributions
AA, PDD, AC applied one or more of the data analytics techniques used in this work. ANC provided data mining guidance and expertise; GPB and SRK provided materials science guidance and expertise. AA led the writing of the paper with inputs from all other authors. All authors read and approved the final submitted manuscript.

Acknowledgements
Tbe authors are grateful to NIMS to make the raw data on fatigue steel strength publicly available. This work is supported in part by the following grants: NSF awards CCF-0833131, CNS-0830927, IIS-0905205, CCF-0938000, CCF-1029166, ACI-1144061, and IIS-1343639; DOE awards DE-FG02-08ER25848, DE-SC0001283, DE-SC0005309, DESC0005340, and DESC0007456; AFOSR award FA9550-12-1-0458.

Author details
[1]Department of Electrical Engineering and Computer Science, Northwestern University, Evanston, IL, USA. [2]Tata Research Development and Design Centre, Tata Consultancy Services, Pune, Maharashtra, India. [3]School of Computational Science and Engineering, Georgia Institute of Technology, Atlanta, GA, USA. [4]Woodruff School of Mechanical Engineering, Georgia Institute of Technology, Atlanta, GA, USA.

References
1.   Committee on Integrated Computational Materials Engineering N. R. C. (2008) Integrated Computational Materials Engineering: A Transformational Discipline for Improved Competitiveness and National Security. http://www.nap.edu/openbook.php?record_id=12199.

2. National Science and Technology Council (2011) Materials genome initiative for global competitiveness. Technical report, National Science and Technology Council. http://www.whitehouse.gov/sites/default/files/microsites/ostp/materials_genome_initiative-final.pdf.

3. Kalidindi SR, Niezgoda SR, Salem AA (2011) Microstructure informatics using higher-order statistics and efficient data-mining protocols. JOM - J Minerals, Met Mater Soc 63(4): 40–41

4. Rajan K (2005) Materials informatics. Materials Today 8(10): 38–45

5. Hey T, Tansley S, Tolle K (2009) The Fourth Paradigm: Data-Intensive Scientific Discovery. Microsoft Research, 1st edition. ISBN: 0982544200, URL: http://research.microsoft.com/en-us/collaboration/fourthparadigm/.

6. Linden G, Smith B, York J (2003) Amazon.com recommendations: item-to-item collaborative filtering. Internet Comput IEEE 7(1): 76–80

7. Mobasher B (2007) Data mining for web personalization. In: Brusilovsky P, Kobsa A, Nejdl W (eds) The adaptive web. Lecture Notes in Computer Science, vol. 4321. Springer-Verlag, Berlin, Heidelberg, pp 90–135

8. Zhou Y, Wilkinson D, Schreiber R, Pan R (2008) Large-scale parallel collaborative filtering for the netflix prize. In: Proceedings of the 4th International Conference on Algorithmic Aspects in Information and Management. AAIM '08, Springer, Berlin, Heidelberg, pp 337–348

9. Das AS, Datar M, Garg A, Rajaram S (2007) Google news personalization: Scalable online collaborative filtering. In: Proceedings of the 16th International Conference on World Wide Web. WWW '07, ACM, New York, NY, USA, pp 271–280

10. URL: Walmart is making big data part of its DNA. Bigdata startups, 2013, http://www.bigdata-startups.com/BigData-startup/walmart-making-big-data-part-dna/.

11. King M (2012) URL: Data Mining the TARGET way. http://www.slideshare.net/ipullrank/datamining-the-target-way.

12. Rajan K, Suh C, Mendez P (2009) Principal component analysis and dimensional analysis as materials informatics tools to reduce dimensionality in materials science and engineering. Stat Anal Data Min 1: 361–371

13. Suh C, Rajan K (2005) Virtual screening and qsar formulations for crystal chemistry. QSAR & Comb. Sci 24(1): 114–119

14. Nowers JR, Broderick SR, Rajan K, Narasimhan B (2007) Combinatorial methods and informatics provide insight into physical properties and structure relationships during ipn formation. Macromol Rapid Commun 28: 972–976

15. Gadzuric S, Suh C, Gaune-Escard M, Rajan K (2006) Extracting information from the molten salt database. Metallogr Mater Trans A 37(12): 3411–3414

16. George L, Hrubiak R, Rajan K, Saxena SK (2009) Principal component analysis on properties of binary and ternary hydrides and a comparison of metal versus metal hydride properties. J Alloys Compounds 478(1–2): 731–735

17. Singh S, Bhadeshia H, MacKay D, Carey H, Martin I (1998) Neural network analysis of steel plate processing. Iron-mak Steelmak 25: 355–365

18. Fujii H, MacKay D, Bhadeshia H (1996) Bayesian neural network analysis of fatigue crack growth rate in nickel base superalloys. ISIJ INT 36: 1373–1382

19. Hancheng Q, Bocai X, Shangzheng L, Fagen W (2002) Fuzzy neural network modeling of material properties. J Mater Process Technol 122(2–3): 196–200

20. Gopalakrishnan K, Ceylan H, Kim S, Khaitan SK (2010) Natural selection of asphalt mix stiffness predictive models with genetic programming. ANNIE Int Eng Syst Artif Neural Netw 20: 10

21. Gopalakrishnan K, Manik A, Khaitan SK (2006) Runway stiffness evaluation using an artificial neural systems approach. Int J Electrical Comput Eng 1(7): 496–502

22. Wen YF, Cai CZ, Liu XH, Pei JF, Zhu XJ, Xiao TT (2009) Corrosion rate prediction of 3c steel under different seawater environment by using support vector regression. Corrosion Sci 51(2): 349–355

23. Rao BV, Gopalakrishna SJ (2009) Hardgrove grindability index prediction using support vector regression. Int J Miner Process 91(1–2): 55–59

24. Gautham BP, Kumar R, Bothra S, Mohapatra G, Kulkarni N, Padmanabhan KA (2011) More Efficient ICME through Materials Informatics and Process Modeling. In: Proceedings of the 1st World Congress on Integrated Computational Materials Engineering (ICME) (eds J. Allison, P. Collins and G. Spanos). John Wiley & Sons, Inc., Hoboken, NJ, USA. doi: 10.1002/9781118147726.ch5

25. Dieter GE (1986) Mechanical Metallurgy. Mc Graw-Hill Book Co. 3rd edition, ISBN: 0-07-016893-8 26. Deshpande PD, Gautham BP, Cecen A, Kalidindi S, Agrawal A, Choudhary (2013) Application of Statistical and Machine Learning Techniques for Correlating Properties to Composition and Manufacturing Processes of Steels. In: 2nd World Congress on Integrated Computational Materials Engineering. John Wiley & Sons, Inc., pp 155-160. ISBN: 9781118767061

26. Deshpande PD, Gautham BP, Cecen A, Kalidindi S, Agrawal A, Choudhary (2013) Application of Statistical and Machine Learning Techniques for Correlating Properties to Composition and Manufacturing Processes of Steels. In: 2nd World Congress on Integrated Computational Materials Engineering. John Wiley & Sons, Inc., pp 155-160. ISBN: 9781118767061

27. URL: National Institute of Materials Science. http://smds.nims.go.jp/fatigue/index_en.html.

28. Weher E, Allen EL (1977) An introduction to linear regression and correlation. (a series of books in psychology.) w. h. freeman and comp., San Francisco 1976. 213 s., tafelanh., s 7.00. Biom J 19(1): 83–84

29. Wang Y (2000) A new approach to fitting linear models in high dimensional spaces. Technical report, University of Waikato, URL: http://books.google.com/books?id=Z0OntgAACAAJ.

30. Wang Y, Witten IH (2002) Modeling for optimal probability prediction. In: Proceedings of the Nineteenth International Conference on Machine Learning. Morgan Kaufmann Publishers, pp 650–657

31. URL: Robust Fit Regression, Mathworks. http://www.mathworks.in/help/stats/robustfit.html.

32. Aha DW, Kibler D (1991) Instance-based learning algorithms. Machine Learning Vol. 6. Kluwer Academic Publishers, Boston, pp 37–66

33. Cleary JG, Trigg LE (1995) K*: An instance-based learner using an entropic distance measure. In: Proceedings of the 12th International Conference on Machine Learning. Morgan Kaufmann Publishers, pp 108–114

34. Kohavi R (1995) The power of decision tables. In: Proceedings of the 8th European Conference on Machine Learning. ECML '95, Springer-Verlag, London, UK, pp 174–189

35. Vapnik VN (1995) The nature of statistical learning theory. Information Science and Statistics Series. Springer-Verlag New York, Inc., New York, NY, USA. ISBN: 0-387-94559-8

36. Bishop C (1995) Neural Networks for Pattern Recognition. 1st edition. Oxford University Press, USA. ISBN: 0198538642

37. Fausett L (1994) Fundamentals of Neural Networks. 1st edition. Prentice Hall, Pearson, New York. ISBN: 0133341860

38. Witten IH, Frank E (2005) Data Mining: Practical Machine Learning Tools and Techniques. The Morgan Kaufmann Series in Data Management Systems, 2nd edition. Morgan Kaufmann Publishers. ISBN: 0120884070

39. Wang Y, Witten IH (1997) Induction of model trees for predicting continuous classes. In: Proc European Conference on Machine Learning Poster Papers, Prague, Czech Republic, pp 128–137

40. Quinlan JR (1992) Learning with continuous classes. In: 5th Australian Joint Conference on Artificial Intelligence. World Scientific, pp 343–348

41. Breiman L, Friedman J, Olshen R, Stone C (1984) Classification and Regression Trees. Wadsworth and Brooks, Monterey, CA

42. R Development Core Team (2011) R: A Language and Environment for Statistical Computing. R Foundation for Statistical Computing, R Foundation for Statistical Computing, Vienna, Austria. ISBN 3-900051-07-0. http://www.R-project.org.

43. MATLAB (2010) Version 7.10.0 (R2010a). The MathWorks Inc., Natick, Massachusetts

44. Hall M, Frank E, Holmes G, Pfahringer B, Reutemann P, Witten IH (2009) The weka data mining software: An update. SIGKDD Explorations Newsletter 11(1): 10-18. doi:10.1145/1656274.1656278, URL: http://doi.acm.org/10.1145/1656274.1656278.

5

OpenCalphad - a free thermodynamic software

Bo Sundman[1*†], Ursula R Kattner[2†], Mauro Palumbo[3] and Suzana G Fries[3]

*Correspondence:
bo.sundman@gmail.com
[†]Equal contributors
[1]INSTN, CEA Saclay, 91191 Saclay, France
Full list of author information is available at the end of the article

Abstract

Thermodynamic data are essential for the understanding, developing, and processing of materials. The CALPHAD (Calculation of Phase Diagrams) technique has made it possible to calculate properties of multicomponent systems using databases of thermodynamic descriptions with models that were assessed from experimental data. A large variety of data, such as phase diagram and solubility data, including consistent thermodynamic values of chemical potentials, enthalpies, entropies, thermal expansions, heats of transformations, and heat capacities, can be obtained from these databases. CALPHAD calculations can be carried out as stand-alone calculations or can be carried out coupled with simulation codes using the result from these calculations as input. A number of CALPHAD software are available for the calculation of properties of multicomponent systems, and the majority are commercial products. The OpenCalphad (OC) software, discussed here, has a simple programming interface to facilitate such integration in application software. This is important for coupling validated thermodynamic as well as kinetic data in such simulations for obtaining realistic results. At present, no other high quality open source software is available for calculations of multicomponent systems using CALPHAD-type models, and it is the goal of the OC source code to fill this gap. The OC software is distributed under a GNU license. The availability of the source code can greatly benefit scientists in academia as well as in industry in the development of new models and assessment of model parameters from both experimental data and data from first principles calculations.

Keywords: Computational thermodynamics; CALPHAD; Software; Multicomponent modeling; Equilibrium calculations; Simulations; GNU license

Introduction

Today's technology relies heavily on materials with complex compositions that often consist of several phases. Proper processing is required to achieve a well-designed microstructure and products with the desired properties. The introduction of new materials, or even minor changes in composition of a known material, often requires a long time for processing adjustments, and development of an entirely new material is even more challenging. The goal of the *Materials Genome Initiative* (MGI) [1], announced in 2011, is to enable the development and deployment of new materials in 'half the time and half the cost'. Computational methods, coupled with traditional experimental methods, are essential for shortening the time-to-market of new products.

Several theoretical methods, ranging from atomistic quantum-mechanics approaches to meso- and macro-scale models, are used together in what is now known as *Integrated Computational Materials Engineering* (ICME) [2]. Computational thermodynamics has been identified as an essential ingredient in ICME and MGI [3] since it can provide input parameters to meso-scale methods such as phase field. Integrating thermodynamic calculations in a simulation of a manufacturing process gives information on the local state of the system, including heat evolution, volume changes, and chemical potentials, that may govern a diffusion process or the movement of a solid/liquid interface. In theory, it is possible to learn from a simulation how to control the external variables to obtain a microstructure with the desired properties of the material. Today, much of this is still done by an expensive trial and error method, relying on the skills and experience of the personnel in charge of the processes.

The CALPHAD (Calculation of Phase Diagrams) method using thermodynamic and other physical property databases is currently the only method available for efficient calculation of the properties of multicomponent, multiphase systems with the accuracy that is required for commercial applications. It has been long recognized [4] that coupling CALPHAD thermodynamics with other physical models for kinetic simulations is useful for better understanding and improvement of many materials processes. As a result, the ability to couple a thermodynamic code with application software is an essential requirement for the realistic modeling of meso-scale simulations within the ICME framework. For example, Olson [5] demonstrated the benefits of this approach for the development of new steels.

In this article, we review the current use of the CALPHAD software, the benefits from the availability of an open source software, and the requirements for use within the ICME framework. We further describe in detail the data structure and formalisms that are used in the OpenCalphad (OC) open source software and comment on databases. We conclude with a summary of the current state of OC and future needs.

Review

The CALPHAD method uses mathematical models to describe the properties of a multicomponent system as a function of temperature, composition, and, if needed, pressure. It was originally developed for describing the thermodynamics of a system for the calculation of phase equilibria but has been extended to calculations of all kinds of equilibria and other thermochemical properties. Experimental data, as well as data from atomistic methods [6], are used for determining the model parameters. The strength of the CALPHAD method is that it provides a consistent and transparent functional description of the properties of the unary, binary, and ternary subsystems that can be combined for the calculation of a multicomponent system. As such, software and databases for CALPHAD calculations have long been utilized by major materials producers for research and development.

Current use of the CALPHAD method

The CALPHAD method was originally developed for the calculation of phase equilibria. As the number of available thermodynamic descriptions for unary, binary, and ternary systems grew, these descriptions were combined into databases for multicomponent systems that were suitable for the calculation of commercial alloys. Initially, CALPHAD calculations were used to determine whether a candidate alloy composition would produce the desired phases and avoid unwanted phases at the temperatures of interest and to select a temperature regime for processing [7]. The combination of

results from CALPHAD calculations with models for other important alloy characteristics, such as creep resistance, castability, and cost, allowed computational testing of
thousands of candidate alloy compositions and the selection of a few promising candidate alloys for experimental evaluation [8]. For finding new candidate alloy compositions,
the CALPHAD calculations can be implemented in systematic searches using, for example, a genetic algorithm [9] or mesh-adaptive direct search algorithm [10]. The remaining
data to fulfill the design criteria can be obtained from artificial neural networks or
Gaussian processes [11]. The materials simulation tool JMatPro [12] uses the results from
a CALPHAD calculation with property models to predict a large number of properties,
including mechanical properties and corrosion behavior [13].

An important aspect of the development of a new material is finding the proper processing conditions to obtain the desired properties. The reliability of the results obtained
from process simulation tools can be greatly improved if these simulations are coupled
with thermodynamic calculations, especially if quantities, such as heat evolution or phase
compositions and phase distributions of a new material, are unlikely to be known from
experiments.

Efficient coupling of the CALPHAD calculations with other software makes a suitable programming interface essential. Several commercial CALPHAD software programs
with interactive user interfaces, such as CaTCalc [14], FactSage [15], MatCalc [16],
MTDATA [17], Pandat [18], and Thermo-Calc [19], are available for the thermodynamic
calculations of alloy systems. Some of these programs have already been coupled by their
developers with diffusion and precipitation simulation software [16,18,20], and most of
these programs have at least one kind of software interface for coupling with application
software.

The idea of linking different length scales through coupling materials science simulation tools is not new [21,22]. CALPHAD calculations have been successfully integrated
with phase field simulations [23], fluid flow simulations [24], and casting simulations [25].
Although thermodynamic calculations have been successfully coupled with phase field
simulations producing realistic microstructures [26], analyses of the mechanical response
of such simulated microstructures are rare at best. It is quite possible that the up-
front investment of having to purchase commercial software licenses may deter many in
academia and small software companies from exploring the feasibility of such coupling.

Benefits from open source code

Commercial casting software today is coupled with commercial CALPHAD software
using a programming interface. Although this path was first paved by coupling with an
older open source code [25], there are many more areas where the software used for materials development and manufacturing would benefit from coupling with thermodynamic
calculations. The same open source code was used in the initial JMatPro software [12].
The benefits of open source codes are well documented [27,28]. Licenses for commercial thermodynamic software are expensive and may not be available for the required
platform. Open source software overcomes these problems and enables researchers and
software developers to explore the benefits from coupling with thermodynamics at
minimal cost.

The approaches used in free software for the calculation of geochemical systems,
GEMS [29] and MELTS [30], are not well suited for the calculation of metallic systems.

Other free software, such as Gibbs [31] or AMPL-based Gibbs energy minimization [32], are either limited to a simple regular solution-type model or have specialized in the mapping of phase diagrams. Although well suited for the calculation of alloy systems, the open source code used by Banerjee et al. [25] lacks the modularity that is needed for the implementation of new models and new kinds of calculations. To date, no general open source software is available for multicomponent CALPHAD calculations. Open source code will make it possible to obtain a better integration between all methods used within the ICME framework, including automating the utilization of data from other methods, such as those from Density Functional Theory (DFT). Open source software will also provide a tool for interested researchers to implement new thermodynamic models, test new algorithms, and extend the functionalities of the software. Open source codes are already available for DFT calculations [33,34], phase field simulations [35,36], and finite element analysis of the mechanical response of microstructures [37]. OC fills the gap that exists between the availability of open source codes for the atomic scale and the microscale.

We present here the first version of a multicomponent open source thermodynamic code distributed under a GNU license, the OpenCalphad (OC) software. This software provides the basic functionalities for the calculation of phase equilibria, phase diagrams, and thermophysical quantities such as chemical potentials, entropies, etc., for stable and metastable states for currently established models.

OC is written in Fortran standard 2008 with extensive use of the data structuring facilities provided by this standard. These data structures are compatible with those of C/C++ and other modern programming languages. As a result, OC can be used on any system for which these compilers are available.

Interfacing application programs with thermodynamics

The data needed from the CALPHAD calculations by application programs are usually phase amounts and compositions, chemical potentials, enthalpies, driving forces, or phase boundary slopes. The application software must be able to set the external conditions for the calculations in a very flexible manner. Transfer of the calculated thermodynamic results to the application software can be made by creating lookup tables for the application software or by direct coupling. The advantage of lookup tables is that their use is frequently faster than direct coupling. However, since lookup tables are generated prior to the execution of the application software, this advantage may come at the expense of accuracy as the conditions during the course of the calculations by the application software may deviate from the original conditions used for the creation of the lookup table.

The programming interfaces of commercial software are black boxes with limited flexibility with respect to hardware and software demands and prescribe which calculations can be performed and which values can be obtained. For the implementation of OC with other application software, it is possible to call the routines of OC directly. Data structure and routines of the General Thermodynamic Package (GTP) are documented in detail, and the documentation is provided together with the source code. Although the direct implementation of the OC routines may be advantageous for reducing execution time of the application software, it can be time consuming to program and debug as special care needs to be taken when directly implementing the routines. The availability of a software interface overcomes this problem. Such a software interface can be a custom-designed interface as was used for the casting simulations by Banerjee at al. [25] or a library of general routines that is provided with the thermodynamic software.

Application software interface

A software interface should consist of a simple set of routines that can select a material from a database; set the external conditions on T, P, and composition; calculate the set of stable phases; and provide the needed property data to the application software.

An established software interface for thermodynamic software called the TQ interface [38,39], originally developed for FactSage [15] and Thermo-Calc [19], has been implemented in OC. This interface provides functions to set the values of external conditions such as T, P, and composition as well as to prescribe the set of stable phases, volume, heat content, etc. of a system and to obtain equilibrium values of phase amounts, phase constitutions, and chemical potentials of all components. The TQ interface can be used to calculate *local equilibria*, such as ongoing transformations during a phase field simulation, and can thus provide important information on thermodynamic and kinetic driving forces. With the TQ interface of OC, OC-TQ, such calculations can be executed in parallel, substantially improving speed.

The TQ interface is an open definition and can be extended to also provide other properties such as mobilities and elastic constants that are needed for the simulation of a phase transition. Such properties often depend on T as well as the phases and their constitutions and can be modeled in a way similar to the thermodynamic properties and stored in databases together with such data.

Implementation of the programming interface

The OC-TQ interface implemented in OC is currently used in the OpenPhase project [35]. The basic OC-TQ interface is written in the new Fortran standard and is extended with a C interface making the routines callable from C++, Java, Python, etc., thus enabling seamless integration into other tools used within the ICME framework, such as phase field [40,41], fluid flow [24], or finite element methods (FEM) [25]. For example, the FEM code used for a casting simulation [25] used two routines, *Lever* and *Slopes*, from a custom interface for an older open source code. These two routines can be replaced by a sequence of routines from the OC-TQ interface, as shown in Table 1. Xiong et al. [42] have provided a number of examples for the implementation of interface routines in kinetic modeling.

Table 1 Conceptual list of essential routines that are available in the OC-TQ interface giving an idea of the functionality

Routine	Action
read_data	Read materials data for a selected set of elements and phases from the database file.
set_condition	Set a condition for T, P, composition, chemical potential, stable phase, etc. Each condition can be set and changed separately. The simulations can use multiple equilibria with different conditions to represent local equilibrium.
calculate_equilibrium	Calculate the equilibrium for the given set of conditions.
get_value	Obtain the calculated value of a chemical potential, amount, or composition of a phase, atomic mobility of a component, etc.

Note that the materials database can also contain data on atomic mobilities, elastic constants, magnetic properties, etc., in addition to the data for the Gibbs energy.

Thermodynamics

A thermodynamic system consists of elements that can form many different molecules or ions, here called species. The elements are neutral monatomic species. The species are the constituents of the phases, and it is important to distinguish between the *constituents* of a phase and the *components*, the latter are usually the same as the elements. The number of constituents can be larger than the number of components; for example, the number of species in the gas phase is frequently larger than the number of elements.

For the calculation of phase equilibria, it is important that the controlling conditions of the system can be specified in a general way. The most simple conditions for minimizing the Gibbs energy are fixed amounts of components, T, and P, but it must be possible to set chemical potentials, values of enthalpy, volume, etc. as well as to prescribe a phase that must be stable. Based on single equilibrium calculations, it must also be possible to generate property diagrams and phase diagrams for varying conditions. The optimal model parameters are determined from experimental and theoretical data in an assessment procedure.

Models and model parameters

All thermodynamic parameters are stored in OC relative to the phases and depend on the model selected for the phase. These parameters can be read from a file (database) containing the phase descriptions. There can be many different kinds of models, from a simple stoichiometric model with no variation in composition to multisublattice models for complex crystalline phases and liquids with a short range order to a gas phase with more than 100 species. The most frequently used models are variants of the Compound Energy Formalism (CEF) [43]. This formalism describes the gas phase, regular solutions, intermetallics, and ordered phases and liquids with a short range order. The general equation for describing the Gibbs energy function per mole formula unit of a phase α is

$$G_m^\alpha = {}^{\text{srf}}G_m^\alpha + {}^{\text{cfg}}G_m^\alpha + {}^{\text{phy}}G_m^\alpha + {}^{E}G_m^\alpha \tag{1}$$

The term ${}^{\text{srf}}G_m^\alpha$ represents the 'surface' of reference for the phase relative to other phases and internal ordering; ${}^{\text{cfg}}G_m^\alpha$ is the configurational term which assumes ideal mixing of the constituents in each sublattice. The term ${}^{\text{phy}}G_m^\alpha$ is used to describe contributions to the Gibbs energy from particular physical phenomena like magnetic transitions, and ${}^{E}G_m^\alpha$ is needed to describe deviations in the Gibbs energy relative to the first three terms. The phase superscript is omitted in the following equations unless several phases are involved.

For example, a phase modeled with two sublattices and two constituents in each is denoted

$$(A, D)_a (B, C)_b \tag{2}$$

where a and b are the stoichiometric ratios. The various terms in Equation 1, using customary CALPHAD notation where the elements or species are separated by ':' if they are on different sublattices and ',' if they are on the same sublattice, for such a model are

$$\text{srf}G_m = y_A^{(1)} y_B^{(2)} {}^\circ G_{A:B} + y_A^{(1)} y_C^{(2)} {}^\circ G_{A:C} + y_D^{(1)} y_B^{(2)} {}^\circ G_{D:B} + y_D^{(1)} y_C^{(2)} {}^\circ G_{D:C} \tag{3}$$

$$^{\text{cfg}}G_m = RT \left(a \left[y_A^{(1)} \ln \left(y_A^{(1)} \right) + y_D^{(1)} \ln \left(y_D^{(1)} \right) \right] + b \left[y_B^{(2)} \ln \left(y_B^{(2)} \right) + y_C^{(2)} \ln \left(y_C^{(2)} \right) \right] \right)$$

$$\text{(4)}$$

$$^{E}G_m = y_A^{(1)} y_D^{(1)} \left(y_B^{(2)} L_{A,D:B} + y_C^{(2)} L_{A,D:C} \right)$$
$$+ y_B^{(2)} y_C^{(2)} \left(y_A^{(1)} L_{A:B,C} + y_D^{(1)} L_{D:B,C} \right) + y_A^{(1)} y_D^{(1)} y_B^{(2)} y_C^{(2)} L_{A,D:B,C}$$

$$\text{(5)}$$

where $y_A^{(1)}$ is the constituent fraction of A in sublattice 1; $^{\circ}G_{A:B}$ is the Gibbs energy of formation of a 'compound' $A_a B_b$, usually called an 'endmember' of the phase; $L_{A,D:B}$ is the interaction parameter between A and D on sublattice 1 when sublattice 2 is filled with B; and R is the molar gas constant. The interaction parameters may depend on T, P, and the constitution. More details about the CEF and other models can be found in Lukas et al. [43].

An efficient and flexible data structure [44] has been implemented in OC to store composition dependencies in the program memory as binary trees as shown in Figure 1. Each binary tree is independent, and this allows parallel processing which can speed up calculations. As shown in the figure, several 'property records' can be linked to each parameter; this allows the description of the composition dependence of other properties other than the Gibbs energy, for example, the Curie temperature and Bohr magneton number for ferromagnetic transitions, mobilities for diffusion data, lattice parameters, etc. This modular data structure is extremely useful for the implementation of new models as new trees can be easily added to the code.

Static and dynamic data

The set of elements, species, and parameters for each phase of a system is 'static' as it does not change with the external conditions. The external conditions set by the user as well as the values of the constituent fractions and the calculated results for each phase are stored in separate 'dynamic' equilibrium records as shown in Figure 2. This approach was chosen to simplify the calculation of several equilibria in parallel, each with its own equilibrium record, for example, for calculating separate lines in a phase diagram or assessing model parameters using many experimental points. The ability for parallel execution of the calculations is especially important for the coupling with other codes, such as phase field, fluid flow, or FEM.

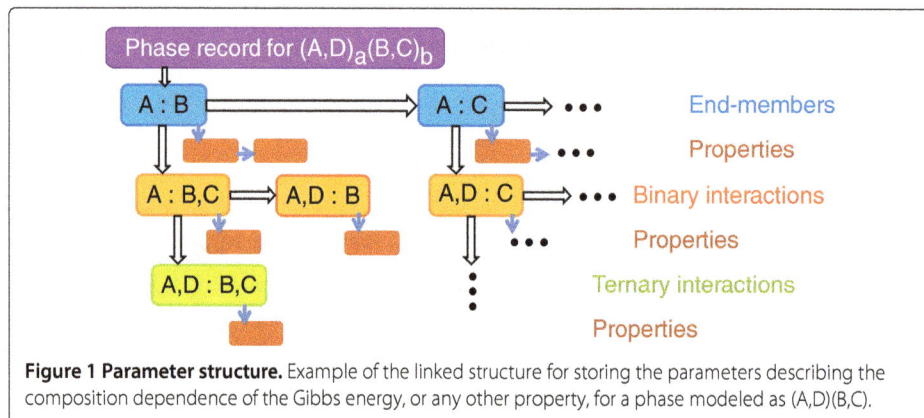

Figure 1 Parameter structure. Example of the linked structure for storing the parameters describing the composition dependence of the Gibbs energy, or any other property, for a phase modeled as (A,D)(B,C).

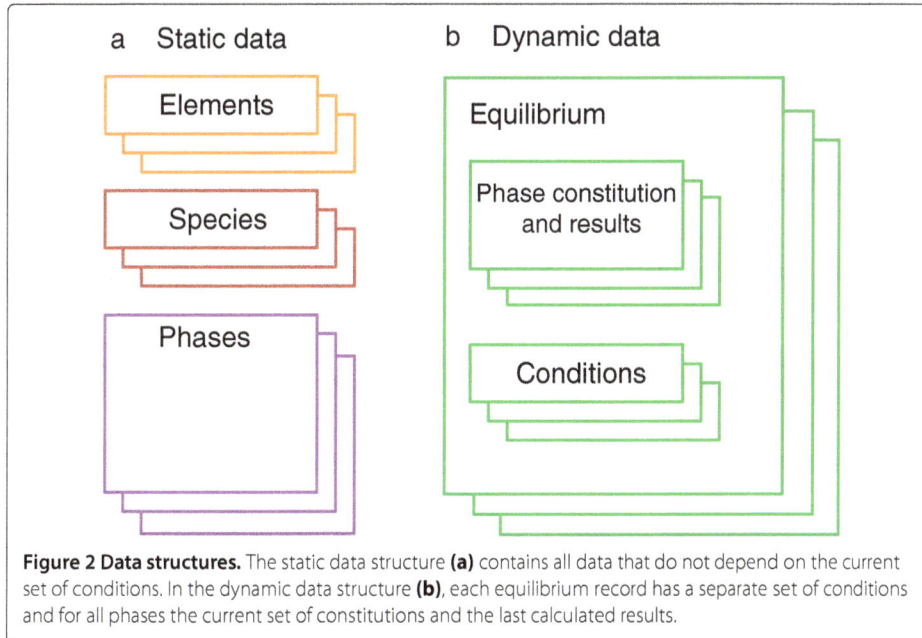

Figure 2 Data structures. The static data structure **(a)** contains all data that do not depend on the current set of conditions. In the dynamic data structure **(b)**, each equilibrium record has a separate set of conditions and for all phases the current set of constitutions and the last calculated results.

Multicomponent equilibrium calculations, the Lagrange method

The total Gibbs energy of the system must be minimized to find the equilibrium at a fixed temperature, pressure, and composition. At the start, neither the set of stable phases nor its constitution is known. The total Gibbs energy is related to the molar Gibbs energy for the current set of stable phases as

$$G = \sum_{\alpha} \aleph^{\alpha} G_m^{\alpha} \tag{6}$$

where $\aleph^{\alpha}$ is the amount of formula units and G_m^{α} the Gibbs energy for one formula unit of the α phase.

The algorithm for minimization implemented in OC was proposed by Hillert [45], and the first implementation was made by Jansson [46]. External conditions and internal constraints are taken into account using Lagrange multipliers. The Lagrange multipliers are also used to replace the Gibbs energy with the appropriate function considering external conditions on the volume, chemical potentials, or prescribed stable phases. An example for the Lagrange function, L, for conditions of constant composition and fixed T and P is

$$L = G + \sum_{A} \left(N_A - \sum_{\alpha} \aleph^{\alpha} M_A^{\alpha} \right) \mu_A + \sum_{\alpha} \sum_{s} \eta^{\alpha,(s)} \left(\sum_{i} y_i^{\alpha,(s)} - 1 \right) \tag{7}$$

where N_A is the total amount of component A, M_A^{α} is the amount of A in phase α, s is the sublattice index, and $y_i^{\alpha,(s)}$ is the constituent fraction of species i on sublattice s in α. μ_A and $\eta^{\alpha,(s)}$ are the Lagrange multipliers. The first constraint is that the mass balance is fulfilled, and the second constraint is that the sum of constituent fractions in each sublattice of α is unity. Hillert [45] has proved that the Lagrange multiplier μ_A is the same as the chemical potential for A. By differentiating the Lagrange function with respect to the different variables many relations can be derived.

The first step of the algorithm

Following Hillert [45], the first step in solving this equation is to invert the matrix of the second derivatives of the Gibbs energy with respect to the constituent fractions:

$$e_{ij}^\alpha = \left(\frac{\partial^2 G_m^\alpha}{\partial y_i \partial y_j} \right)^{-1} \tag{8}$$

These matrix elements are used in several equations in the second step of the algorithm. Hillert [45] has shown that the Gibbs-Duhem relation can be used for the following equation to correct the constituent fractions, Δy_i^α, in the phase α at each iteration:

$$\Delta y_i^\alpha = \sum_A \sum_j \frac{\partial M_A^\alpha}{\partial y_j^\alpha} e_{ij}^\alpha \mu_A - \sum_j \frac{\partial G_m^\alpha}{\partial y_j^\alpha} e_{ij}^\alpha - \sum_j \frac{\partial^2 G_m^\alpha}{\partial T \partial y_j^\alpha} e_{ij}^\alpha \Delta T \tag{9}$$

where μ_A is the chemical potential of component A, and ΔT is the change in temperature at the iteration unless T is constant. All partial derivatives are calculated analytically in OC. But to calculate Δy_i^α from Equation 9, we must first find μ_A and ΔT which is done in a second step. A term similar to that for ΔT is introduced into Equation 9 for ΔP if P is variable.

The second step of the algorithm

In the second step, we formulate a system of linear equations based on the external conditions in order to obtain a new set of values of the chemical potentials, amounts of the stable phases, and temperature and/or pressure if they are variable. There is one equation for each stable phase:

$$\sum_A M_A^\alpha \mu_A = G_m^\alpha \tag{10}$$

if any chemical potential is fixed, this term is moved to the right hand side. For each condition on an extensive variable, we obtain an equation which in a simple case where N_A is fixed is

$$\sum_\alpha \aleph^\alpha \sum_i \frac{\partial M_A^\alpha}{\partial y_i^\alpha} \sum_B \sum_j \frac{\partial M_B^\alpha}{\partial y_j^\alpha} e_{ij}^\alpha \mu_B + \sum_\alpha M_A^\alpha \Delta\aleph^\alpha$$
$$= \sum_\alpha \aleph^\alpha \sum_i \sum_j \frac{\partial M_A^\alpha}{\partial y_i^\alpha} \frac{\partial G_m^\alpha}{\partial y_j^\alpha} e_{ij}^\alpha \tag{11}$$

where we find the matrix elements e_{ij}^α again.

The new values of the chemical potentials allow the updating of the phase constitutions using Equation 9. If the amount of a phase becomes negative, the phase should be removed from the set of stable phases, and if the driving force, ΔG, for an unstable phase β

$$\Delta G^\beta = \sum_A M_A^\beta \mu_A - G_m^\beta \tag{12}$$

becomes positive, this phase is added to the set of stable phases. Equation 12 requires that the constitution of the metastable phases be updated using Equation 9 at each iteration.

When the changes in chemical potentials and phase amounts are sufficiently small and all external conditions are fulfilled, the calculation has converged; otherwise, the calculation is repeated beginning with the first step.

A variation of this algorithm has been proposed by Lukas [47] in which the two steps were combined and variables like Δy_i^α are found directly by solving a larger system of linear equations than is needed in step 2 above. This is more efficient if the set of stable phases does not change between the iterations, but each time such a change occurs, the matrix of linear equations must be reconstructed. The two-step method does not require such reconstruction.

Calculations using the same models and parameters and the same external conditions with different software will give the same results if the criteria for convergence are similar. The results from the present code agree with the results from other codes, such as PMLFKT [47]. Rounding errors are the likely cause for differences in the sixth or seventh significant digit of the results.

Start values, the grid minimizer

The algorithm described above is an iterative method which depends strongly on the initial configuration of the phases. In Figure 3, the Gibbs energy curves for two phases are shown. If the start constitutions of the two phases are as marked in Figure 3b, an iterative algorithm will find the common tangent shown in Figure 3c. This is a metastable equilibrium because it lies above the most stable common tangent as shown in Figure 3d.

To avoid the problem of not finding the global equilibrium for the given conditions, Chen et al. [48,49] developed an approach for finding the stable equilibria and values for the start constitutions of the phases. Subsequently, similar approaches were implemented into other software such as Thermo-Calc [19] or CaTCalc [14].

The OC software has as an initial step a grid minimizer which calculates the Gibbs energy of all phases over a grid of compositions as shown in Figure 4a. These grid points are treated as stoichiometric phases, illustrated in Figure 4b. In a preliminary minimization, the set of grid points representing the minimum as shown in Figure 4c is determined. The grid points from the grid minimizer are then used as an initial guess, and their compositions are inserted as the start constitutions for the iterative algorithm which will find the global minimum in Figure 4d. Even in multicomponent systems, only a few 1,000 grid points are needed for each phase.

Databases

Thermodynamic descriptions are read by OC from thermodynamic database (TDB) files [50]. This format has developed into a *de facto* standard and is supported by most commercial software. However, the OC code will not be able to read proprietary,

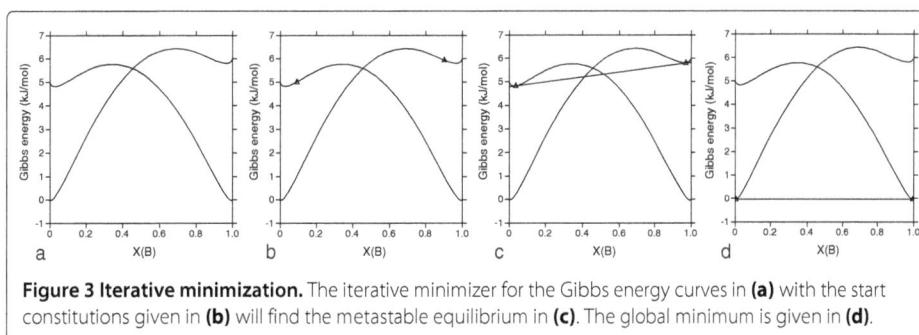

Figure 3 Iterative minimization. The iterative minimizer for the Gibbs energy curves in (**a**) with the start constitutions given in (**b**) will find the metastable equilibrium in (**c**). The global minimum is given in (**d**).

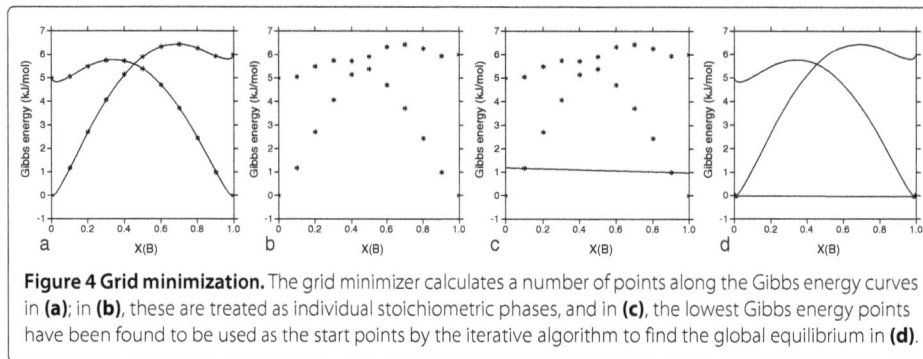

Figure 4 Grid minimization. The grid minimizer calculates a number of points along the Gibbs energy curves in **(a)**; in **(b)**, these are treated as individual stoichiometric phases, and in **(c)**, the lowest Gibbs energy points have been found to be used as the start points by the iterative algorithm to find the global equilibrium in **(d)**.

encrypted database files. Files with descriptions of individual systems or databases for multicomponents can be found as supplemental material of journal articles or in file repositories [51].

It should be noted here that uncertainty of results from the CALPHAD calculations depends on the databases used and how reliably the original experimental data used in the development of these databases are reproduced. Currently, no method for uncertainty quantification of the results from a CALPHAD database is available. The reliability of the results is most commonly expressed by plotting experimental data versus calculated results under the same conditions [52]. Recently, a systematic evaluation of the effect of using different DFT sets of data to compute phase diagrams and thermodynamic properties has been carried out [53,54]. The work of Palumbo et al. [53] showed that differences of a few kJ/mol in the input DFT energies of formation can produce topologically different phase diagrams with phases that appear/disappear as stable ones depending on the input data set used. Similar effects can be obtained using different sets of optimized parameters from experiments (and eventually DFT data) obtained by the CALPHAD approach. Mathematical methodologies to better evaluate such uncertainties are desirable but challenging (if at all possible). The OC software offers a useful tool for future implementations of such methods by scientific communities different from the traditional users (physicists or materials scientists).

Conclusions

The benefits from using the results from the CALPHAD calculations for materials design and process development are well established. We have highlighted the many pathways by which the results from the CALPHAD calculations can be utilized within the ICME framework.

The properties that are currently implemented by the CALPHAD software and databases are the Gibbs energy, molar volume, diffusion, and precipitation kinetics. However, there are many other physical properties that are phase based, and the approach used by the CALPHAD method is well suited for modeling them [55]. The description of these properties will be most likely tied to the thermodynamic description of the individual phases and for expanding the CALPHAD code to include these properties; the availability of a source code for the calculation of thermodynamics will greatly benefit the efficiency of the new code. Since the inception of the CALPHAD method, models for temperature, pressure, and compositions have been evolving to better reflect the underlying physics of the phases. For example, new models for the description of the unaries are being discussed [56-60] and expansion of the existing thermodynamic code may be necessary. Such developments are only possible for those who have access to the source code of commercial software unless open source code is available.

In this article, we have presented OC, an open source code for the calculation of phase equilibria and thermodynamic properties. The OC software has currently the general CEF implemented but more models can be added by interested users. A well-established algorithm has been selected for finding the equilibrium. This algorithm can handle different types of conditions, including the specification of phases which should be stable, and has a powerful method for providing start values to ensure that global minimum is obtained. The minimization algorithm can handle changes in the set of stable phases and detect miscibility gaps.

Current state of OC

The development of a free thermodynamic software for academic research is underway, and a first version of the OC software is available [61]. The present version provides the basic functionalities necessary for multicomponent thermodynamic calculations. Currently, implemented features in the downloadable version are

- The standard multicomponent CEF model with multiple sublattices including magnetism, regular solution, and gas. Composition-dependent binary and ternary parameters are also implemented.
- Command line user interface with macro facilities.
- Interactive inputing, editing and listing of models and model parameters.
- Ability to read models and model parameters from a standard, unencrypted, thermodynamic databases.
- Ability to set conditions for T, P, amounts or fractions of components, chemical potentials, stable phases.
- Multicomponent equilibrium calculations finding the set of stable phases, including phases with miscibility gaps, without any the need for initial estimates.
- Extensive documentation and some test cases as macro files.
- Parameters for volumes, thermal expansion, and bulk modulus.

Implemented but not yet released are

- The ionic liquid model [62] for liquids with short range ordering.
- A tentative software interface to ICME [2] applications written in C++.
- Storage of non-thermodynamic data that can depend on T, P, and composition [55], such as mobilities, elastic constants, electrical resistivity, and viscosity.
- STEP and MAP procedures; the simple step procedure, as shown in Figure 5, is already implemented.

Plans for future features include

- New models for unaries for proper utilisation of results from DFT calculations [33,34,56-60].
- Module for the assessment of model parameters.
- Implementation of a corrected quasichemical model for liquid [63].

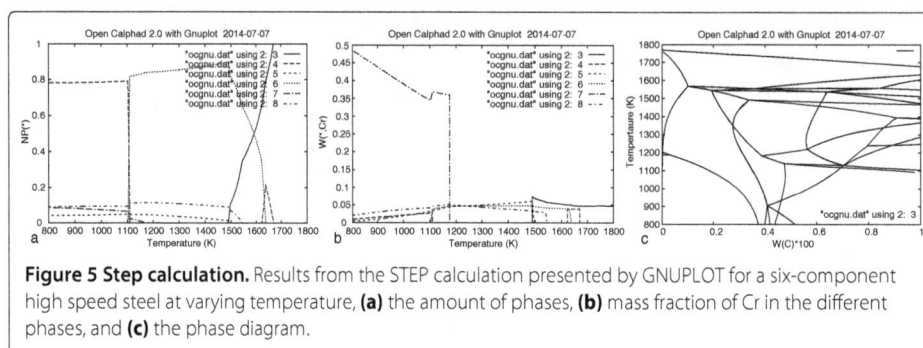

Figure 5 Step calculation. Results from the STEP calculation presented by GNUPLOT for a six-component high speed steel at varying temperature, **(a)** the amount of phases, **(b)** mass fraction of Cr in the different phases, and **(c)** the phase diagram.

Interested readers are encouraged to download the current version [61] and give feedback to the developers or participate in further development. It is important that the developers receive feedback from users who are willing to take the time to test this version.

Desirable future developments beyond open source code

It was previously mentioned that no method for uncertainty quantification of the results from the CALPHAD calculations is currently available. The first logical step in this direction is to perform a sensitivity analysis of the model parameters that are being adjusted during the optimization of the thermodynamic description of a binary or ternary system.

In addition to uncertainty quantification, it is desirable to develop a set of test cases to ensure that the results are consistent with accepted codes. Although the results from calculations should agree once the convergence criteria are fulfilled, it is not always guaranteed that the true set of equilibrium phases has been found. Although grid minimizers are intended to assure that the true equilibrium is found, the reliability of these results will depend on how the grid was established.

The development of tools for uncertainty quantification, verification, and validation requires a transparent code that can be easily modified to accommodate the required tests. The availability of an open source code is essential to address these needs and opens the door for future development and improvement of the CALPHAD method and for further integration with materials and process simulation tools.

Competing interests
The authors declare that they have no competing interests.

Authors' contributions
BS and URK contributed the majority to the present article. MP and SGF contributed in the formalism development and discussed and contributed to the present article. All authors read and approved the final manuscript.

Acknowledgements
Funding by the Interdisciplinary Centre for Advanced Materials Simulation (ICAMS), which is supported by ThyssenKrupp AG, Bayer MaterialScience AG, Salzgitter Mannesmann Forschung GmbH, Robert Bosch GmbH, Benteler Stahl/Rohr GmbH, Bayer Technology Services GmbH, and the state of North-RhineWestphalia as well as the European Commission in the framework of the European Regional Development Fund (ERDF). One of the developers, BS, is grateful for a Humboldt senior research award.
NIST does not endorse any commercial products, and use of these products does not imply endorsement by NIST.

Author details
[1]INSTN, CEA Saclay, 91191 Saclay, France. [2]Materials Science and Engineering Division, National Institute of Standards and Technology, 100 Bureau Dr., Gaithersburg, MD 20899, USA. [3]ICAMS, Ruhr University Bochum, Universitätsstr. 150, 44780 Bochum, Germany.

References

1. National Science and Technology Council (2011) Materials genome initiative for global competitiveness, Office of Science and Technology Policy, Washington, DC
2. National Research Council, Committee on Integrated Computational Materials Engineering (2008) Integrated computational materials engineering: a transformational discipline for improved competitiveness and national security, National Research Council, Committee on Integrated Computational Materials Engineering, Washington, DC
3. Olson GB (2014) Preface to the viewpoint set on: The Materials Genome. Scripta Mater 70:1–2
4. Kattner UR, Eriksson G, Hahn I, Schmid-Fetzer R, Sundman B, Swamy V, Kussmaul A, Spencer PJ, Anderson TJ, Chart TG, Costa e Silva A, Jansson B, Lee B-J, Schalin M (2000) Use of thermodynamic software in process modelling and new applications of thermodynamic calculations. Calphad 24:55–94
5. Olson G (2013) Genomic materials design: The ferrous frontier. Acta Mater 61:771–781
6. Hickel T, Kattner UR, Fries SG (2014) Computational thermodynamics: Recent developments and future potential and prospects. Phys Status Solidi B 251:9–13
7. Olson GB (1997) Computational design of hierarchically structured materials. Science 277:1237–1242
8. Reed RC, Tao T, Warken N (2009) Alloys-by-design: application to nickel-based single crystal superalloys. Acta Mater 57:5898–5913
9. Xu W, Rivera-Díaz-del-Castillo PEJ, van der Zwaag S (2008) Genetic alloy design based on thermodynamics and kinetics. Phil Mag 88:1825–1833
10. Gheribi AE, Robelin C, Le Digabel S, Audet C, Pelton AD (2011) Calculating all local minima on liquidus surfaces using the FactSage software and databases and the mash adaptive direct search algorithm. J Chem Thermodyn 43:1323–1330
11. Tancret F (2013) Computational thermodynamics, Gaussian processes and genetic algorithms: combined tools to design new alloys. Modelling Simul Mater Sci Eng 21:045013
12. Saunders N, Kucherenko S, Li X, Miodownik AP, Schille J-P (2001) A new computer program for predicting materials properties. J Phase Equilib 22:463–469
13. JMatPro Practical Software for Materials Properties. http://www.sentesoftware.co.uk/jmatpro.aspx. Accessed 19 December 2014
14. Shobu K (2009) CaTCalc: New thermodynamic equilibrium calculation software. Calphad 33:279–287
15. Bale CW, Bélisle E, Chartrand P, Decterov SA, Eriksson G, Hack K, Jung I-H, Kang Y-B, Melançon J, Pelton AD, Robelin C, Petersen S (2009) FactSage thermochemical software and databases - recent developments. Calphad 33:295–311
16. MatCalc The Materials Calculator. http://matcalc.tuwien.ac.at/. Accessed 19 December 2014
17. Davies RH, Dinsdale AT, Gisby JA, Robinson JAJ, Martin SM (2002) MTDATA - thermodynamic and phase equilibrium software from the national physical laboratory. Calphad 26:229–271
18. Cao W, Chen S-L, Zhang F, Wu K, Yang Y, Chang YA, Schmid-Fetzer R, Oates WA (2009) The Pandat software with PanEngine, PanOptimizer and PanPrecipitation for multi-component phase diagram calculation and materials property simulation. Calphad 33:328–342
19. Andersson J-O, Helander T, Höglund L, Shi P, Sundman B (2002) Thermo-Calc & DICTRA, computational tools for materials science. Calphad 26:273–312
20. Shi P, Engström A, Sundman B, Ågren J (2011) Thermodynamic calculations and kinetic simulations of some advanced materials. Mater Sci Forum 675-677:961–974
21. Liu Z-K, Chen L-Q, Raghavan P, Du Q, Sofo JO, Langer SA, Wolverton C (2004) An integrated framework for multi-scale materials simulations and design. J Comput-Aided Mater Des 11:183–199
22. Liu Z-K, Chen L-Q, Krishna R (2006) Linking length scales via materials informatics. JOM 58(11):42-50
23. Grafe U, Böttger B, Tiaden J, Fries SG (2000) Coupling of multicomponent thermodynamic databases to a phase field model: application to solidification and solid state transformations of superalloys. Scripta Mater 42:1179–1186
24. Schneider MC, Gu JP, Beckermann C, Boettinger WJ, Kattner UR (1997) Modeling of micro- and macrosegregation and freckle formation in single-crystal nickel-base superalloy directional solidification. Metall Mater Trans A 28A:1517–1531
25. Banerjee DK, Samonds MT, Kattner UR, Boettinger WJ (1997) Coupling of phase diagram calculations for muliticomponent alloys with solidification micromodels is casting simulation software. In: Beech J, Jones H (eds). Solidification Processing 1997: Proceedings of the 4th Decennial International Conference on Solidification Processing, Department of Engineering Materials, University of Sheffield, Sheffield. pp 354–357
26. Warnken N, Ma D, Dreverman A, Reed RC, Fries SG, Steinbach I (1997) Phase-field modeling of as-cast microstructure evolution in nickel-based superalloys. Acta Mater 57:5862–5875
27. 10 Reasons Open Source Is Good for Business. http://www.pcworld.com/article/209891/. Accessed 19 December 2014
28. Ince DC, Hatton L, Graham-Cumming J (2012) The case for open computer programs. Nature 482:485–488
29. GEM Software (GEMS) Home. http://gems.web.psi.ch. Accessed 19 December 2014
30. MELTS Home Page. http://melts.ofm-research.org/. Accessed 8 December 2014
31. Cool T, Bartol A, Kasenga M, Modi K, García RE (2010) Gibbs: Phase equilibria and symbolic computation of thermodynamic properties. Calphad 34:393–404
32. Snider J, Griva I, Sun X, Emelianenko M (2015) Set based framework for Gibbs energy minimization. Calphad 48:18–26
33. Gonze X, Beuken J-M, Caracas R, Detraux F, Fuchs M, Rignanese G-M, Sindic L, Verstraete M, Zerah G, Jollet F, Torrent M, Roy A, Mikami M, Ghosez P, Raty J-Y, Allan DC (2002) First-principles computation of material properties: the ABINIT software project. Comput Mater Sci 25:478–489
34. Giannozzi P, Baroni S, Bonini N, Calandra M, Car R, Cavazzoni C, Ceresoli D, Chiarotti GL, Cococcioni M, Dabo I, Dal Corso A, de Gironcoli S, Fabris S, Fratesi G, Gebauer R, Gerstmann U, Gougoussis C, Kokalj A, Lazzeri M, Martin-Samos L, Marzari N, Mauri F, Mazzarello R, Paolini S, Pasquarello A, Paulatto L, Sbraccia C, Scandolo S,

Sclauzero G, Seitsonen AP, *et al* (2009) Quantum ESPRESSO: a modular and open-source software project for quantum simulations of materials. J Phys: Condens Matter 21:395502

35. OpenPhase. http://www.openphase.de/. Accessed 19 December 2014
36. Guyer JE, Wheeler D, Warren JA (2009) FiPy: partial differential equations with Python. Comput Sci Eng 11(3):6–15
37. Langer SA, Fuller Jr ER, Carter WC (2001) OOF: an image-based finite-element analysis of materials microstructures. Comput Sci Eng 3(3):15–23
38. Eriksson G, Sippola H, Sundman B (1994) A proposal for a general thermodynamic calculation interface. In: Jokilaakso A (ed). 1st Colloquium on Process Simulation. Helsinki University of Technology, Helsinki. pp 67–103
39. TQ-Interface Programmer's Guide. http://www.thermocalc.com/support/documentation/. Accessed 19 December 2014
40. Steinbach I, Böttger B, Eike J, Warnken N, Fries SG (2007) CALPHAD and phase-field modeling: a successful liaison. J Phase Equilib Diffus 28:101–106
41. Kitashima T (2008) Coupling of the phase-field and CALPHAD methods for predicting multicomponent, solid-state phase transformations. Phil Mag 88:1615–1637
42. Xiong H, Huang Z, Wu Z, Conway PP (2011) A generalized computational interface for combined thermodynamic and kinetic modeling. Calphad 35:391–395
43. Lukas HL, Fries SG, Sundman B (2007) Computational Thermodynamics, the Calphad Method. Cambridge Univ. Press, Cambridge
44. Sundman B (1981) Application of computer techniques on the treatment of the thermodynamics of alloys. PhD thesis, KTH Stockholm, Sweden
45. Hillert M (1981) Some viewpoints on the use of a computer for calculating phase diagrams. Physica 103B:31–40
46. Jansson B (1984) Computer operated methods for equilibrium calculations and evaluation of thermochemical model parameters, PhD thesis, KTH Stockholm, Sweden
47. Lukas HL, Weiss J, Henig E-T (1982) Strategies for the calculation of phase diagrams. Calphad 6:229–251
48. Chen S-L, Chou K-C, Chang YA (1993) On a new strategy for phase diagram calculation. 1. basic principles. Calphad 17:237–250
49. Chen S-L, Chou K-C, Chang YA (1993) On a new strategy for phase diagram calculation. 2. binary systems. Calphad 17:287–302
50. Thermo-Calc Database Manager's Guide. http://www.thermocalc.com/support/documentation/. Accessed 19 December 2014
51. NIST Repositories. http://nist.matdl.org/, https://materialsdata.nist.gov/dspace/xmlui. Accessed 19 December 2014
52. Bratberg J, Mao H, Kjellqvist L, Engström A, Mason P, Chen Q (2012) The development and validation of a new thermodynamic database for Ni-based alloys. In: Huron ES, Reed RC, Hardy MC, Mills MJ, Montero RE, Portella PD, Telesman J (eds). Superalloys 2012: 12Th International Symposium on Superalloys. TMS (The Minerals, Metals & Materials Society), Warrendale, PA. pp 803–812
53. Palumbo M, Fries SG, Hammerschmidt T, Abe T, Crivello J-C, Al Hasan Breidi A, Joubert J-M, Drautz R (2014) First-principles based phase diagrams and thermodynamic properties of tcp phases in Re-X systems (X = Ta, V, W). Comput Mater Sci 81:433–445
54. Lejaeghere K, Van Speybroeck V, Van Oost G, Cottenier S (2014) Error estimates for solid-state density-functional theory predictions: an overview by means of the ground-state elemental crystals. Crit Rev Solid State Mater Sci 39:1–24
55. Campbell CE, Kattner UR, Liu Z-K (2014) The development of phase-based property data using the CALPHAD method and infrastructure needs. Integr Mater Manuf Innov 3:12
56. Palumbo M, Burton B, Costa e Silva A, Fultz B, Grabowski B, Grimvall G, Hallstedt B, Hellman O, Lindahl B, Schneider A, Turchi PEA, Xiong W (2014) Thermodynamic modelling of crystalline unary phases. Phys Status Solidi B 251:14–32
57. Becker CA, Ågren J, Baricco M, Chen Q, Decterov SA, Kattner UR, Perepezko JH, Pottlacher GR, Selleby M (2014) Thermodynamic modelling of liquids: CALPHAD approaches and contributions from statistical physics. Phys Status Solidi B 251:33–52
58. Körmann F, Al Hasan Breidi A, Dudarev SL, Dupin N, Ghosh G, Hickel T, Korzhavyi P, Muñoz JA, Ohnuma I (2014) Lambda transitions in materials science: recent advances in CALPHAD and first-principles modelling. Phys Status Solidi B 251:53–80
59. Hammerschmidt T, Abrikosov IA, Alfè D, Fries SG, Höglund L, Jacobs MHG, Koßmann J, Lu X-G, Paul G (2014) Including the effects of pressure and stress in thermodynamic functions. Phys Status Solidi B 251:81–96
60. Rogal J, Divinski SV, Finnis MW, Glensk A, Neugebauer J, Perepezko JH, Schuwalow S, Sluiter MHF, Sundman B (2014) Perspectives on point defect thermodynamics. Phys. Status Solidi B 251:96–129
61. OpenCalphad. http://www.opencalphad.org/, https://github.com/sundmanbo/opencalphad. Accessed 19 December 2014
62. Hillert M, Jansson B, Sundman B, Ågren J (1985) A two-sublattice model for molten solutions with different tendency for ionization. Metall Trans A 16A:261–266
63. Hillert M, Selleby M, Sundman B (2009) An attempt to correct the quasichemical model. Acta Mater 57:5237–5244

6

The Penn State-Georgia Tech CCMD: ushering in the ICME Era

Zi-Kui Liu[1] and David L McDowell[2*]

* Correspondence:
david.mcdowell@me.gatech.edu
[2]Woodruff School of Mechanical
Engineering, School of Materials
Science and Engineering, Georgia
Institute of Technology, Atlanta, GA
30332-0405, USA
Full list of author information is
available at the end of the article

Abstract

This case study paper presents the origins, philosophy, organization, development, and contributions of the joint Penn State-Georgia Tech Center for Computational Materials Design (CCMD), a NSF Industry/University Cooperative Research Center (I/UCRC) founded in 2005. As a predecessor of and catalyst for Integrated Computational Materials Engineering (ICME), the CCMD served as a basis for coupling industry, academia, and government in advancing the state of computational materials science and mechanics across a portfolio of process-structure-property-performance relations, with emphasis on education and training of the future workforce in computational materials design.

Keywords: ICME; MGI; CCMD; NSF I/UCRC; Materials design; Computational materials science

Background

The past decade has witnessed the emergence of *Integrated Computational Materials Engineering* (ICME) as an early twenty-first century joint industry, academic, and government initiative for integration of modeling and simulation with materials development and product improvement. ICME is concerned with multiple levels of structure hierarchy, as is typical of materials, and aims to reduce the time to market of innovative products by exploiting concurrent design and development of materials, products, and process/manufacturing paths. As described in the National Materials Advisory Board committee [1] report from the National Research Council, ICME is 'an approach to design products, the materials that comprise them, and their associated materials processing methods by linking materials models at multiple length scales'. ICME embraces the engineering perspective of a top-down, goal-means strategy discussed cogently by Olson [2] and is fully cognizant of the important role of microstructure in tailoring materials properties/responses in most engineering applications, well beyond the atomic or molecular scale. Many materials properties/responses not only depend on atomic bonding and atomic/molecular structure but are also strongly influenced by the existence, spatial arrangement, and morphology of multiple phases and resulting phase interface/interphase strengthening effects.

This perspective was embraced and refined for the academic, industry, and government research communities at a 1998 National Science Foundation (NSF)-sponsored workshop hosted by Georgia Tech and Morehouse College [3] entitled 'New Directions in Materials

Design Science and Engineering (MDS&E)'. As stated in the workshop report, 'The field of materials design is entrepreneurial in nature, similar to such areas as microelectronic devices or software. MDS&E may very well spawn a "cottage industry" specializing in tailoring materials for function, depending on how responsive large supplier industries can be to this demand. In fact, this is already underway'. That workshop report concluded that a change of culture is necessary in U.S. universities and industries to cultivate and develop the concepts of simulation-based design of materials to support integrated design of materials and products. It also forecasted that the twenty-first century global economy would usher in a revolution of the materials supply/development industry and realization of true virtual manufacturing capabilities, not only geometric modeling but also consideration of realistic material behavior. It was recommended to establish a national roadmap addressing (i) databases for enabling materials design, (ii) developing principles of systems design and the prospects for hierarchical materials systems, and (iii) identifying opportunities and deficiencies in science-based modeling, simulation, and characterization 'tools' to support concurrent design of materials and products.

Inspired by the 1998 NSF MDS&E workshop report and the educational effort at Northwestern University led by Olson, Zi-Kui Liu, Long-Qing Chen, and Karl Spear at Penn State spearheaded development of a fundamentally new kind of computational materials science curriculum and laboratory experience in 2001 to 2003 [4,5]. This NSF-funded effort established a computational teaching facility and addressed theoretical and computational aspects of thermodynamics and kinetics. Thermodynamic and kinetic databases in the forms of Gibbs energy functions and atomic mobility in individual phases were utilized through computer programs to predict phase stability and simulate phase transformations [6]. As shown in Figure 1, thermodynamics is at the core of this conceptual framework to establish configurations, either stable or metastable, and driving forces for microstructure evolution. Then, considering kinetics of transition states for defects and crystallography, properties and performance (structure-property relations) can be modeled. Experiments play a vital role in motivating, calibrating, and validating models at various time and length scales.

This educational activity further inspired them to develop the Materials Computation and Simulation Environment (MatCASE) program [4,5] in 2002, and in the same year, the name MaterialsGenome® was coined by Professor Liu in establishing the

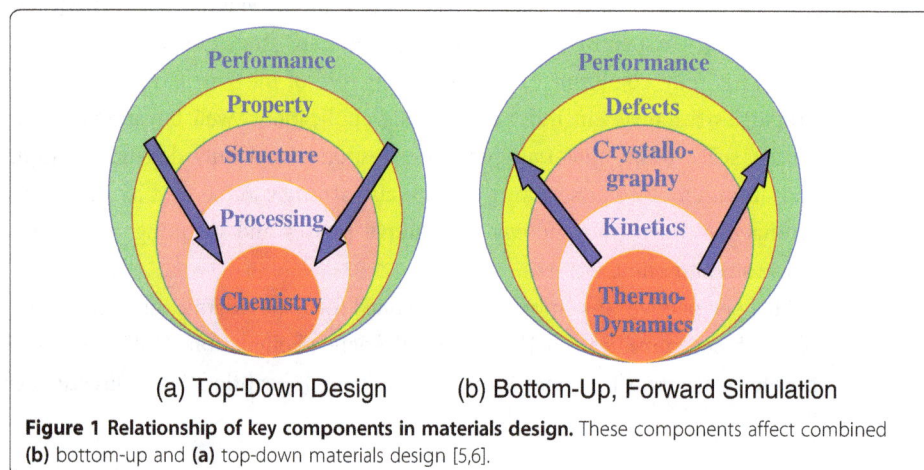

(a) Top-Down Design   (b) Bottom-Up, Forward Simulation

Figure 1 Relationship of key components in materials design. These components affect combined **(b)** bottom-up and **(a)** top-down materials design [5,6].

MaterialsGenome, Inc. in Pennsylvania and later trademarked (cited by the MGI web site at http://www.whitehouse.gov/mgi). In the MatCASE program, the Penn State team developed unique strengths in conducting multilevel modeling by passing information from first-principles calculations and computer coupling of phase diagrams and thermochemistry (calculation of phase diagrams (CALPHAD)) modeling to phase-field simulations and structure-property relations, which was later leveraged into the Center for Computational Materials Design (CCMD). It was realized that first-principles calculations based on density functional theory (DFT) were becoming a critical component in not only providing insights to physics of phenomena but also quantitatively predicting thermodynamic, kinetic, and mechanical properties of individual phases, thus significantly enhancing the predictability of CALPHAD modeling of individual phases in multicomponent materials [6,7]. The properties of individual phases and phase interfaces thus obtained are used as input parameters for phase-field simulations of microstructure evolutions. The microstructures resulting from phase-field simulations, along with properties of phases and phase interfaces, enter into the finite element analysis of materials responses to external stimuli. Thus, in the hierarchy of material structure [8-10], individual phases can be considered as the building blocks of materials in designing microstructures that meet desired performance requirements.

Prior to forming the CCMD, the 2001 to 2003 Defense Advanced Research Projects Agency (DARPA)-funded *Accelerated Insertion of Materials* (AIM) program [11-13] sought to build systems approaches to accelerate the insertion of new and/or improved materials into products. The AIM program demonstrated that legacy materials development (both polymer composites and aircraft gas turbine materials) could be significantly accelerated by integrating process-structure and structure-property modeling with processing and experiments using a designer knowledge database in collaborative teams involving the materials supply chain, original equipment manufacturers (OEMs), and university research laboratories.

Case description

Forming the CCMD

This historical setting formed the basis for initial discussions starting in 2003 between Zi-Kui Liu and David McDowell in framing a computational materials design initiative within the context of a NSF Industry/University Cooperative Research Center (I/UCRC). At that time, there was no I/UCRC in existence or other federally funded center that dealt with the materials process-structure-property-performance paradigm shown in Figure 1. The Penn State-Georgia Tech collaboration was conceived to take advantage of the pioneering work in computational thermodynamics and phase-field modeling at Penn State, complementing recognized leadership in experimental and computational microstructure-property relations and systems-based materials design methods at Georgia Tech.

The initial conceptualization of the CCMD considered materials design objectives, with a primary goal to characterize sensitivity of properties to microstructure and process route and to capture essential dominant mechanisms and their transitions with applied loading and environment in applications. A related challenge is addressing the uncertainty associated with forms of models, model parameters, microstructure stochasticity, and microstructure hierarchy, in addition to the transfer of information from fine grain, high-resolution models to coarse grain models with reduced degrees of freedom at higher scales. Given these sources of uncertainty, the notion of design *optimization* using hierarchical or concurrent multiscale models is not particularly useful in many cases. Instead, extension of concepts of systems-based *robust* design to multilevel integrated design of materials and products [8-10] is more practical, with sensitivity of various responses to microstructure variation playing a central role. A combined top-down and bottom-up strategy that served to inspire the CCMD is shown in Figure 2, with application to Ni-base superalloys for aircraft gas turbine engines.

This conceptual basis for a center that would develop novel predictive algorithms and methods to support materials design and development led to engagement of a set of initial industry and government stakeholders in fall 2003 to write a letter expressing support for the concept of the CCMD, including Air Products, Inc., ALCOA, Allegheny Ludlum, Boeing Company, Caterpillar Inc., ExxonMobil Upstream Research Company, Ford Motor Company, GE Global Research Center, GE Power Systems, General Motors, Honeywell-Aerospace, Intel, KennaMetal, Inc., Marlow Industries, Inc., Nippon Steel Corporation (Japan), Pratt & Whitney, Questek LLC, RTI International, Special Metals Company, ThermoCalc Software, AB (Sweden), Timken, Los Alamos National Laboratory, Lawrence Livermore National Lab, NASA, Natural Resources Canada (Canada), NIST, Oak Ridge National Lab, Sandia National Labs, SRI International, Exponent, GE Aircraft Engines, DuPont, Argonne National Laboratories, Synaps, Inc., Dana Technology Development Group, U.S. Air Force Research Laboratory (AFMC), and SI Flooring Systems. This led to submission of a planning proposal to the NSF I/UCRC program to establish the CCMD in 2004, followed by a 21 January 2005 planning workshop held at Penn State. The strategy to form the CCMD rested on the complementary nature of strengths. Penn State's world class capabilities in computational thermodynamics and phase-field theory were

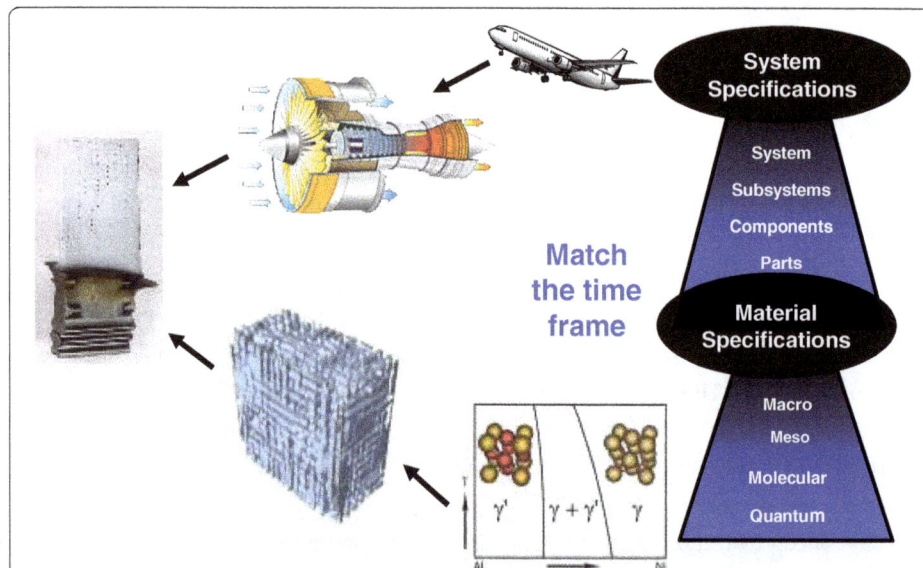

Figure 2 Combined bottom-up and top-down strategy for concurrent design of materials and systems.
Engineering systems design should be extended to the material as a sub-system, with hierarchy of atomistic, mesoscopic, and macroscopic scales, in this case applicable to gamma-prime precipitate strengthened Ni-base superalloys used in the hot section of aircraft gas turbine engines.

combined with Georgia Tech's widely recognized expertise in microstructure-property relations and systems-based materials design. Based on the discussion and feedback from the industry and government labs, a full proposal was submitted to the NSF in June 2005. The proposal for phase I CCMD was funded, with the 1.5-day kickoff meeting held on 3 to 4 November 2005 at Penn State, with 13 initial members. Phase I was in effect from 2005 to 2010. A follow-on phase II was funded by the NSF from 2010 to 2013. This time frame spans the era preceding and leading up to the National Materials Advisory Board report on ICME, as well as the 2011 launch of the U.S. Materials Genome Initiative (MGI) [14].

CCMD vision, mission, structure, and operations

The mission and vision of the CCMD remained consistent over its entire duration of NSF funding:

Mission: Educate the next generation of scientists and engineers with a broad, industrially relevant perspective on engineering research and practice.

Vision: To be recognized as the premier entity for collaborative activities in computational materials design among universities, industries and government laboratories.

The intellectual merit of the CCMD was based on the integration of multiscale, interdisciplinary computational expertise at Penn State and Georgia Tech, ranging from atomistic calculations to continuum phase-field, finite element, and statistical continuum microstructure-property modeling with interfaces between engineering systems design, information technology, and physics-based simulation of process-structure and structure-property relations of materials. Details of membership and research projects are herein protected from disclosure owing to the terms of the Memorandum of Agreement (MOA). Suffice it to say that the CCMD provided leadership in articulating the importance of integrated design of materials and products to industry and the broad profession of materials engineering and developed a significant body of new methods for estimating structure and corresponding properties/responses based on first-principles, atomistic, phase-field, and finite element strategies. Moreover, a component of phase I contributions added value to algorithms for concurrent design of components and materials and decision-based design methods. Tools and methods explored by the CCMD included first-principles calculations, CALPHAD, phase-field, crystal plasticity, molecular dynamics, cohesive finite element methods, homogenization, and systems integration and design tools. Materials systems addressed include Al, Ni, Ti, Mg, and Nb alloys, as well as steels.

Professor Zi-Kui Liu at Penn State served as CCMD Director and was responsible for center activities. He managed membership dues contributions and allocations to funded projects at both Penn State and Georgia Tech, based on input from the Member Advisory Board (MAB) management team, and tracked status of dues collections and new member recruiting, assisted by Center Manager at Penn State, Sandy Watson. In addition to communication among PIs at both universities, students, members, and potential members, Penn State maintained the CCMD website (http://www.ccmd.psu.edu/).

Professor Dave McDowell served as Co-Director of the CCMD and directed the Georgia Tech site. He collaborated closely with Professor Liu in all aspects of assessing progress on projects, developing and pursuing the vision for the CCMD, member recruiting, retention, planning meetings, interactions, and monitoring mentoring relationships that members offer to CCMD-supported students and was assisted at Georgia Tech by Cecelia Jones in organizing annual CCMD meetings in Atlanta.

The management structure of the CCMD is shown in Figure 3. The CCMD had a MAB, in lieu of the Industry Advisory Board (IAB) label used by many I/UCRCs, comprised of one representative per member organization, with different numbers of votes for full and associate members. The CCMD management team was comprised of the CCMD Director, Co-Director, MAB Chair and Vice-Chair, and members for the respective University Policy Committees (Penn State and Georgia Tech administrative representatives). The independent Center Evaluator applied online assessment tools ('Level of Interest and Feedback Evaluation' forms) at every CCMD meeting for each project presented, attended closed MAB meetings, provided liaison with members to discuss any concerns with CCMD management or policies, and administered annual member and faculty surveys to acquire feedback regarding the overall progress of the Center. Faculty, students, and postdocs at both universities interfaced with the MAB, Center Evaluator, and CCMD Management Team, as shown in Figure 3.

The CCMD represented a ground-breaking effort to instill the culture shift associated with ICME, viewing materials design as an integral part of multidisciplinary engineering systems design. Both Professors Liu and McDowell maintained heavy involvement in external workshops and conference presentations, often presented jointly, to publicize the ICME-oriented perspective of the CCMD and the field of computational materials design to the external community [15-28]. With The Minerals, Metals, and Materials Society (TMS) as a partner organization, CCMD management was intimately involved with offering presentations and workshops at TMS meetings.

Development of partnerships among industry, academia, and government laboratories was emphasized through:

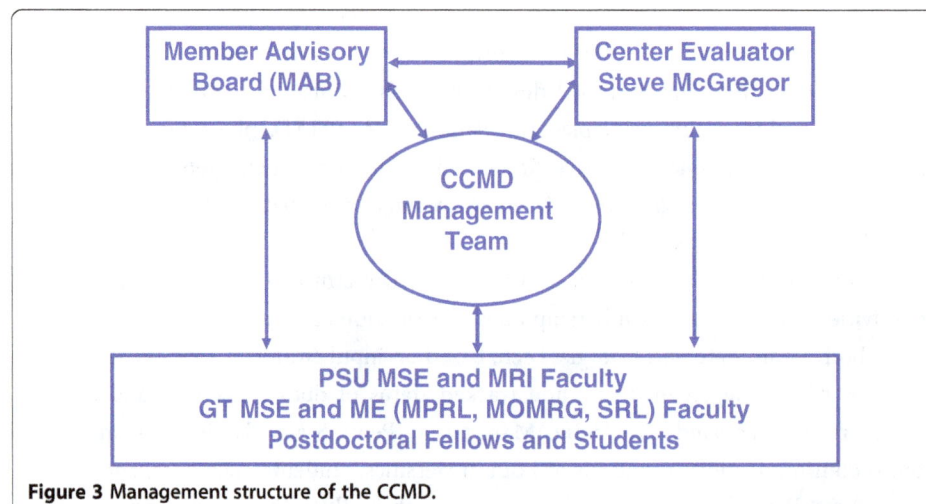

Figure 3 Management structure of the CCMD.

- Educating future generations of scientists and engineers in ICME/MGI pertinent research themes;
- Improving the intellectual capacity of the workforce through industrial participation, high-quality research projects in computational materials science and materials design;
- Promoting research programs of interest to both industry and academia;
- Enhancing the infrastructure of computational materials research in the nation; and
- Exploring and extending physics-based simulations of process-structure and structure-property relations of materials.

Working with CCMD membership, final forms of its MOA and Bylaws were established during the early years of the CCMD. The MOA addressed center objectives, member advisory board, reports including invention disclosures and patent protection and patent rights, royalties derived from licensing, rights in software, data, and publications, confidential communications, publicity, supersedure, representation, termination, indemnity, satellite sites, and warranty disclaimer. The CCMD Bylaws governed the operation of the Center, including membership qualification, privilege, benefits, revocation, and costs, and procedures for proposal voting, project funding, and project reporting.

The CCMD established policies in its MOA for sharing intellectual property developed by funded projects among center members and has successfully implemented such sharing via license agreements. These policies have served the CCMD well in terms of intellectual property policies in accordance with the Bayh-Dole Act, as policies for distribution of software and codes developed in the CCMD are clearly set forth that permit non-exclusive, royalty-free licenses for center members and the possibility of exclusive, royalty-bearing licenses. Although provision for patents was made available as part of the MOA to membership, no such patents were pursued, as is typical for I/UCRCs, and the CCMD effectively operated as a pre-competitive research consortium.

Industry and national laboratory member dues provided the primary financial resources for the CCMD. The CCMD maintained a two-tiered membership structure, $40 K (phase I)/$48 K (phase II) per year for full members and $15 K (phase I)/$18 K (phase II) per year for associate (SBIR eligible small company) members. A very strong incentive for members to join the CCMD was provided by the NSF stipulation that the universities should charge only 10% F&A (overhead) on research projects funded by member dues (Penn State provided further cost sharing with additional overhead reductions). Annual NSF funding of the CCMD was used to partially offset administrative operations (Penn State MRI provided additional support annually). A significant fraction of NSF funding was used to support the Independent NSF Center Evaluator, who served a role as liaison between center membership, leadership, and the NSF, particularly during phase II in which the full cost for such compensation was covered by Penn State, consisting of more than half of the NSF support provided.

Clearly, the level of per project funding was sufficient to support the primary objective of preparing the future workforce in computational materials design, creating an interface between students and postdocs and stakeholder companies and laboratories interested in ICME. During this period, the CCMD was arguably well ahead of the curve in producing such students relative to other academic programs, with 41 graduate students and postdoctoral fellows fully or partially supported. One student was supported by a DoD laboratory internship through completion of his doctoral degree in summer 2010, more than 2 years beyond the end of the formal CCMD funding. A number of students obtained summer internships at member organizations over the years. Important goals for student development were set as follows:

Leadership experience:

- Introduction to industry applications
- Participation and presentations at bi-annual reviews and meetings
- Competition among students for R&D funding and communicating ideas
- Group/team work - collaborate on and between projects

Networking:

- Networking and contacts with industry, academic, and government members
- Guidance and direction via interaction with industry, university, and government
- Lab tours and workshops - communicate with visitors/members

Industry experience:

- Exposure to industry applications and culture - budgets, timelines, competitors, IP
- Research proposals - how to develop 'fundable' ideas
- Feedback from industry sponsors - learn what is important to industry
- Mentoring - project and career guidance
- Internship and employment opportunities

CCMD meetings, available for participation only by members and potential members as per NSF I/UCRC guidelines, were held twice per year for 1.5 days each, including a mid-February meeting at Georgia Tech and a mid-August meeting at Penn State. Prior to each August meeting, project ideas were solicited from members, with sufficient lead time for faculty to prepare proposals. These proposals were distributed to members approximately 2 to 4 weeks before the meeting and were presented by faculty at the meeting. Thereafter, the projects were discussed at length and ranked by members. Based on this input, new projects were then finalized in the following 1 to 2 months by the Chair and Vice-Chair of the Member Advisory Board and CCMD Director and Co-Director, taking into account balance of the portfolio shown in Figure 4, as well as distribution between Penn State and Georgia Tech sites in accordance with attribution of membership dues recruiting. In addition, several industry members hosted the CCMD meetings. The typical agenda for the August meetings consisted of presenting proposals for the next round of funded projects in response to member-developed initiatives offered in late spring or early summer of each year. The February meetings at Georgia Tech focused on student presentations of progress on active CCMD projects, receiving guidance and feedback for the next 6 months. Poster sessions and workshops were commonly held 1 day in advance of each meeting, affording members an opportunity to delve into greater detail and learn modeling principles, interface with CCMD software and algorithms, and explore applicability to their organizations.

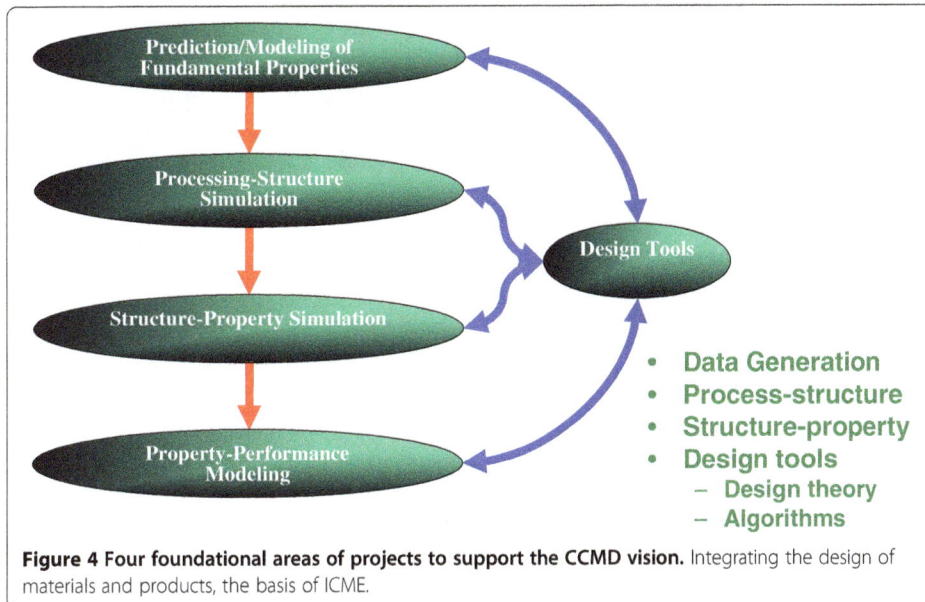

Figure 4 Four foundational areas of projects to support the CCMD vision. Integrating the design of materials and products, the basis of ICME.

At any point in time, given the membership levels, approximately ten projects were underway simultaneously. Projects were configured for a 2-year period, with an option to apply for renewal that was encouraged only for projects receiving strong support and feedback from membership. Projects in phase I focused on filling out a balanced portfolio of research and development in the four foundational elements shown in Figure 4. Materials of focus in phase I included FCC Al and Ni-base alloys, HCP Mg and Ti-alloys, steels, ceramics/oxides, and polymers.

Based on feedback from the Member Advisory Board in 2008, monthly web meetings were organized with presentations for projects led by supported graduate students and postdocs. In addition to monthly web-based presentations by students, mentor-led meetings were organized by supported students. Project mentors from the CCMD membership offered guidance through the year and in some cases provided additional support for on-site student internships. During phases I and II, a total of 38 CCMD projects were executed. In addition to the associated tools and methods, member benefits from CCMD projects included:

- Active interactions with many faculty members with different expertise
- Influence on pre-competitive CCMD projects
- Contributions to education through mentoring projects
- Networking with other CCMD members

The business model for flow of value from CCMD efforts in developing novel modeling and simulation tools to enable computational materials design is shown in Figure 5. It served as the approach for transition from basic cross-cutting research in the CCMD to applications involving specific alloy systems or other materials of proprietary interest to members. CCMD projects supported innovative, publishable basic research that fostered development of graduate students, while the transition to industry-specific applications was funneled downstream into internships, active mentoring roles, and additional research contracts between CCMD members and Penn State or Georgia Tech faculty.

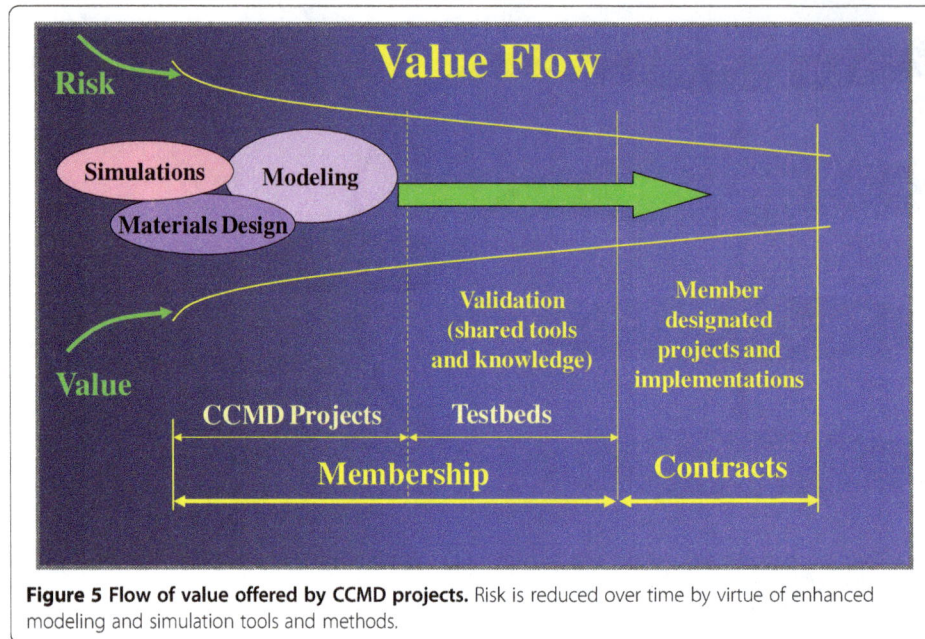

Figure 5 Flow of value offered by CCMD projects. Risk is reduced over time by virtue of enhanced modeling and simulation tools and methods.

The primary research findings were summarized in the project deliverables report submitted with each annual report. The members-only portion of the CCMD website (http://www.ccmd.psu.edu/) posted project quarterly reports and presentations, papers, the project final report, project deliverables, and associated documentation, providing ease of access for members. Moreover, updated quad charts were provided for each project to assist members in communicating relevance of CCMD accomplishments and deliverables within their organizations.

In addition to Center memberships, TMS joined as a partner organization of the CCMD in 2006, with an agreement to publicize the CCMD and host workshops. The CCMD co-organized the annual ASM-TMS Symposium on Computational Materials Design in 2007, held at GE Global Research Center. A workshop, 'Center for Computational Materials Design: Experiences & Perspectives Workshop', was held in 2009 at the Materials Science and Technology (MS&T) conference with partial support from Office of International Business Development, Department of Community & Economic Development, Pennsylvania's Center for Trade Development. The CCMD contributed significantly to establishing the annual symposium at the MS&T conference: Phase Stability, Diffusion Kinetics, and Their Applications (PSDK), which was initiated in 2006.

In May 2007, the CCMD teamed with the Center for Dielectric Study (CDS) at Penn State to submit a successful proposal on 'Computational Modeling of Defects and Minor Chemical Additives in Functional Materials' to the TIE program at NSF, which sought to link efforts of multiple I/UCRCs. This work focused on the thermodynamics and defects formation in perovskites, starting from the prediction of properties of constituent pure element and binary systems such as Ti, TiO_2, and $PbTiO_3$. A 2-year NSF supplement funding project IIP-0823907, *Fundamental Supplement Proposal: Bridging First-principles and Molecular Dynamics Methods to Support Alloy Design in the CCMD*, was funded from 2008 to 2010. Investigators included D.L. McDowell, T. Zhu, and K. Jacob from Georgia Tech and Z.-K. Liu and V. Crespi from Penn State. By definition, simulation-based materials design requires computational exploration of new materials that have not previously been envisioned or developed. This necessitates the use of first-principles and atomistic simulations to estimate fundamental properties of crystals and phases, thereby facilitating consideration in design.

From 2010 to 2013 in phase II, the CCMD focused more on interfaces between phases in addition to fundamental phase properties, congruent with the phase II vision shown in Figure 6 outlined in the renewal proposal. Specific additional gaps addressed in phase II of the CCMD are outlined in the vision shown in Figure 6:

- Linkage of first-principles calculations to higher scales of hierarchy in structure-property simulations, e.g., linking Penn State models, codes, and expertise in the former to Georgia Tech models, codes, and expertise in the latter.
- Mapping modeling, simulation, and design tools developed within the CCMD to various material classes and application domains.
- Materials processing experiments and simulations.

The CCMD made key advances in setting the tone for collaboration and future work-force training that addressed the academic-industry-government cultural paradigm shift towards computationally assisted materials design and development associated with ICME [1] and the more recently framed MGI [14]. In addition to research, outreach was an important element of CCMD operations. In 2007, the CCMD participated in the Women in Science and Engineering Research (WISER) program at Penn State. In this program, female freshmen start their research activities in their second semester at Penn State. The CCMD also explored various NSF supplementary support including Research Experience for Undergraduates (REU), Research Experience for Teachers (RET), and Research Experience for Veterans (REV). In one exemplary success story from the RET program, a math teacher and a group of students from local high school near Penn State worked with faculty and graduate students of the CCMD for several

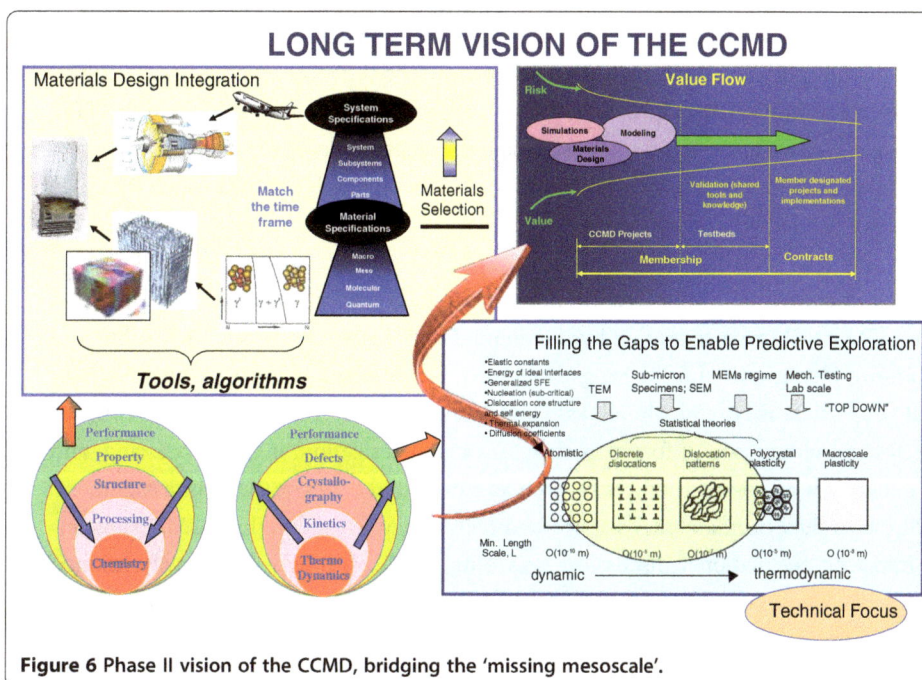

Figure 6 Phase II vision of the CCMD, bridging the 'missing mesoscale'.

years on several projects related to data analysis and geometry. The math teacher and his students participated and presented their posters at CCMD meetings. The REU/RET/REV and WISER programs at Penn State were complemented with programs at Georgia Tech, including the NSF-funded Summer Undergraduate Research Fellowship (SURF) program and the Georgia Industrial Fellowship for Teachers (GIFT) program, which coordinates recruitment of high school STEM teachers for summer positions in research laboratories.

CCMD accomplishments and impact

The CCMD spawned fundamentally new science and technology developments in support of ICME. CCMD projects pioneered the following revolutionary new scientific and engineering in terms of methods and tools to support systems-based computational materials design:

- First-principles prediction methods for elastic constants, thermal expansion coefficient, and antiphase boundary energies (Liu, PSU).
- Framework for automation of CALPHAD thermodynamic modeling (Liu, PSU).
- Software-engineered phase-field codes and libraries to facilitate parametric studies of grain growth and coarsening phenomena (Chen, PSU).
- Methodology for complete treatment of nucleation phenomena using diffuse interface phase-field models (Du and Chen, PSU).
- Quantitative prediction of plane strain fracture toughness of realistic microstructures using cohesive zone models (Zhou, GT).
- Extreme value statistics approaches for high-cycle fatigue strength informed by multiple computational realizations of polycrystals (McDowell, GT).
- Computational methods for effects of inclusions on fatigue strength/life of high-strength bearing steels in rolling and sliding contact (Neu, GT).
- Linkage of phase-field model predictions for polycrystalline structures with continuum polycrystal plasticity simulations to close the loop on process-structure-property relations (Chen, PSU, in collaboration with Garmestani and McDowell, GT).
- Comprehensive approach for robust design of materials based on hierarchical computational modeling and simulation, with a monograph published in 2009 by Elsevier (McDowell, Mistree, and Allen, GT).

A range of codes and tools were developed and made available to members as listed on the CCMD web page accessible by members. For example, Professor Zi-Kui Liu and Drs. ShunLi Shang and Huazhi Fang developed a web-based tool to calculate fundamental properties of Ni and Mg alloys. Graduate student Yan Li, advised by Professor Min Zhou, developed a GUI for microstructure characterization, image processing, and mesh generation of microstructures, as well as automated cohesive element assignment. In addition to these fundamental science and technology advances, the CCMD substantially impacted the national discussion regarding ICME, through the aforementioned presentations at major conferences of materials societies (e.g., [15-28]) as well as book chapters, conference proceedings, and archival journal articles related directly to ICME challenges (e.g., [29]).

CCMD members have provided input regarding the impact of the CCMD and the vision of Integrated Computational Materials Engineering within their organizations. For example, Figures 7 and 8 present CCMD member testimonials of impact on their organizations that were presented and submitted to the NSF. The U.S. Army Research Laboratory reported that their organization 'is actively pursuing the materials-by-design approach to accelerate/shorten the materials development time line. As such, independent of the CCMD, but essentially with the same philosophy, we have been hiring more modelers along all of the appropriate skill set being able to deal with different modeling length scales'.

CCMD faculty members were successful in developing grant proposals beyond internal CCMD projects, written in collaboration with CCMD members or in part due to their affiliation with the Center, or written as an outcome of faculty liaison with previous or current members. Some of these projects were in the form of direct funding from current or previous CCMD members, and others were subcontracts from current or previous CCMD members as flow through on federal prime funding to the members. Over \$3 M total awarded proposal budgets were reported for 'CCMD-spinoff' research, representing additional funding arising from relationships with the CCMD members.

ICME relevant contributions to education

Two new courses were approved in 2007 and are now part of regular courses in the materials science major and the computational science minor at Penn State. Both courses have been further updated with results from CCMD research activities. The two courses are taught in alternative years:

- MatSE580: Computational Materials Science I: Computational Thermodynamics
- MatSE581: Computational Materials Science II: Continuum, Mesoscale Simulations

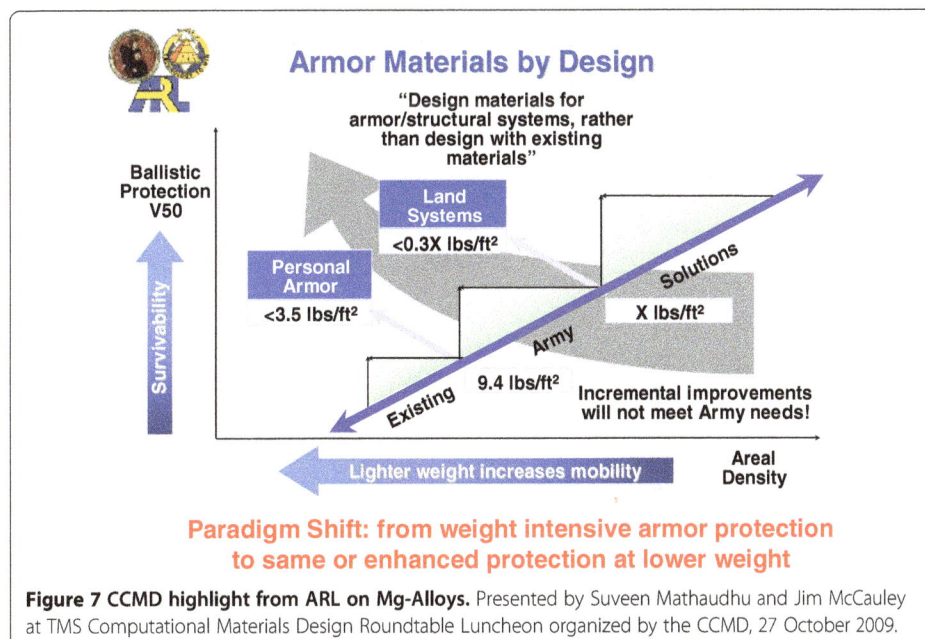

Figure 7 CCMD highlight from ARL on Mg-Alloys. Presented by Suveen Mathaudhu and Jim McCauley at TMS Computational Materials Design Roundtable Luncheon organized by the CCMD, 27 October 2009.

Simulation Tool for Hertzian Roller Contact Fatigue Modeling

Tapered roller bearing assembly

The Timken Company is dedicated to finding solutions to improve its customers' performance. Timken Engineers and Scientists have a history of employing models to accelerate innovations in their products and processes. The Center for Computational Materials Design (CCMD) provides Timken with access to world leading investigators and to the very latest techniques for understanding and modeling microstructural evolution and mechanical response of materials. Many of the projects within the CCMD are positioned to have a positive impact on Timken's ability to provide solutions for their customers.

Case Study - Timken produces steel and roller bearings from steel. Because bearings are designed to have long operational lives, life testing of a new bearing process or a new bearing design can take months of run time on roller bearing fatigue testing machines. These long test times necessarily require careful and limited selection of variants which in turn can result in missed opportunities. The process of making and testing prototypes can cost 10's or 100's of thousands of dollars, and can be a hurdle in generating new business.

Early stage of a raceway spall

Timken and the CCMD have partnered to generate a physically based simulation tool to represent the initiation of hertzian rolling contact fatigue damage as a function of rolling contact parameters, microstructure, and hard oxide inclusion shape, size and orientation. This model employs a crystal plasticity approach as a physically based representation of plastic damage accumulation. Coupled with additional modeling and simulations tools available at Timken, this crystal plasticity model is being used to interrogate broad design options in steel and bearing manufacturing in order to achieve advanced designs and shorter lead times.

Inclusions

Fatigue Cracks   25 μm

Once fully realized, these capabilities represent potential savings in design and test cost and generation of new business valued at several million dollars per year.

Figure 8 CCMD highlight from Timken Company on rolling contact fatigue.

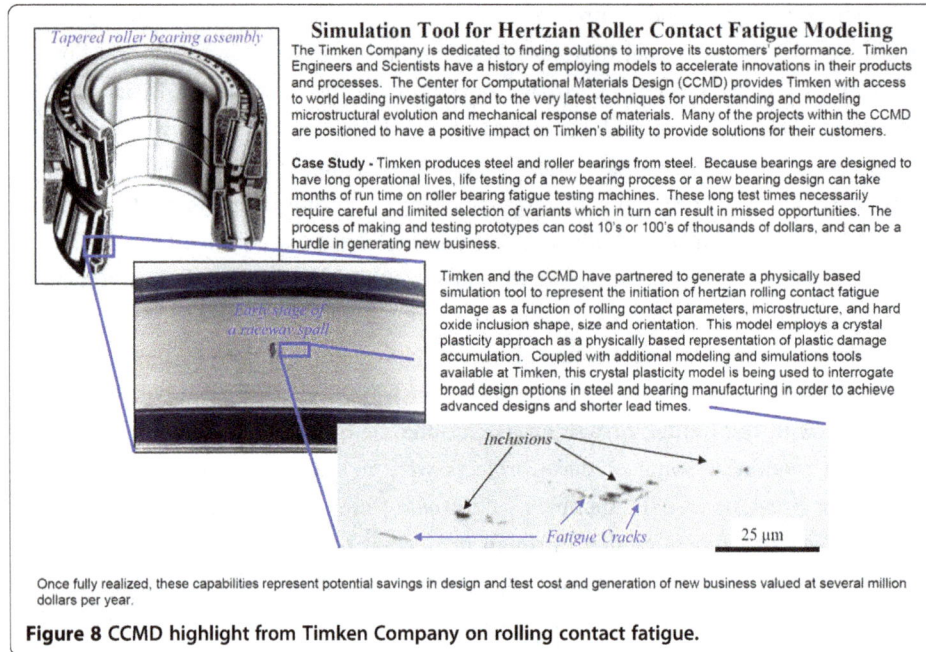

Professors Liu and Chen at Penn State have integrated computational components in their respective undergraduate and graduate courses: Thermodynamics of Materials (401), Phase Relations (410), Thermodynamics of Materials (501), and Kinetics of Materials Processing (503).

Professsor McDowell developed a new advanced graduate course at Georgia Tech ME/BME 7205 entitled 'Mechanics and Applications of Nanostructured Materials and Devices.' The course is team-taught and covers quantum mechanics, molecular modeling and applications in mechanical and thermal properties and responses of interest. Elements of materials design were added to ME4213, Materials Selection & Failure Analysis, a technical elective primarily aimed at seniors at Georgia Tech and led by Professor R.W. Neu. McDowell also introduced a new junior level required core course in Materials Science and Engineering (MSE 3025, *Statistics and Numerical Methods in Materials Science and Engineering*) that is consistent with trends in Integrated Computational Materials Design and the CCMD vision of greater awareness and capability of undergraduate students in materials science and engineering in the areas of statistical methods and numerical methods/computational materials science. Several graduate level courses in a materials design sequence at Georgia Tech have been positively influenced by the CCMD, including:

- MSE 6795 Mathematical, Statistical and Computational Techniques in Materials Science;
- MSE 6796 Structure-Properties Relationships in Materials; and
- MSE 6797 Thermodynamics and Kinetics of Microstructural Evolution

McDowell served as lead-author on one of the first ICME-relevant textbooks [10] entitled 'Integrated Design of Multiscale, Multifunctional Materials and Products', a collaborative effort that had its roots in the CCMD and other ICME-relevant research programs. The layout of chapters in this book shown in Table 1 reflects the multidisciplinary nature of ICME; it is broader than just a computational materials science-centered activity, showing that materials design and development is an integrated activity involving designers, materials suppliers, OEMs, characterization and testing labs, and manufacturers. The linkage to products is clear from the title as well, reflecting the top-down, requirements driven nature of materials development for economic and societal benefits. This is also a clear underlying theme of the Materials Genome Initiative [14].

Clearly, such a book could not be written from the singular perspective of a materials scientist, chemist, or physicist, nor by a designer, nor by a manufacturing specialist, etc. ICME textbooks that aim to prepare the future workforce will need to enrich the linkages and bring fields together. We have learned from our CCMD experience that ICME requires large-scale integration of stakeholders in an innovation ecosystem.

Discussion and evaluation

The CCMD was originally envisioned to embrace the intersection of materials modeling and simulation at various length and time scales with systems design. One may consider such a goal as pertaining to integrated materials design and development problems with a focus on a given material class and under constraint of limited resources. In reality, the diversity of member interests in various distinct materials classes in the CCMD consortium made it difficult to focus on a specific material system or design/development application. It became apparent that limited available resources from member dues within such a consortium drove projects towards addressing specific gaps in modeling and simulation, rather than overall frameworks for integrating process-structure-property-performance relations.

Some lessons learned related to the capability of the structure of the NSF I/UCRC program to support ICME-related research. The prospect of leveraging core NSF funding was helpful in attracting industry membership in the CCMD. In particular, the NSF I/UCRC program requirement of a limitation to 10% indirect costs associated with projects funded by member dues served as a significant incentive to join the CCMD rather than fund university research independently at much higher rates. However, as ICME

Table 1 Chapter titles for 'Integrated Design of Multiscale, Multifunctional Materials and Products' [10]

Chapter numbers	Chapter titles
1	Integrated Material, Product and Process Design - A New Frontier in Engineering Systems Design
2	Critical Path Issues in Materials Design
3	Overview of the Framework for Integrated Design of Materials, Products and Design Processes
4	Decision-Making in Engineering Design
5	Mathematical Tools for Decision-Making in Design
6	Robust Design of Materials - Design Under Uncertainty
7	Integrated Design of Materials and Products: Robust Topology Design of a Cellular Material
8	Integrated Design of Materials and Products: Robust Design Methods for Multilevel Systems
9	Concurrent Design of Materials and Products: Managing Design Complexity
10	Distributed Collaborative Design Frameworks
11	Closure: Advancing the Vision of Integrated Design of Materials and Products

developed, industry appeared to be interested in leveraging higher levels of federal investment provided by other sources (e.g., DoD, DoE) than the NSF I/UCRC program. The NSF funding level in the I/UCRC ($50 K per year to Georgia Tech and $60 K to lead institution Penn State during phase I, reduced in phase II) was too low to justify pursuit of cohesive foundational engineering problems or to entice member companies to explore extensions of their in-house proprietary materials design and development protocols. Moreover, a significant fraction of the NSF budget supported the Independent NSF Evaluator and I/UCRC mechanisms for running meetings (e.g., real-time web-based feedback from members during presentations); it was not clear that these expenditures and modes for evaluations/discussion provided higher utility in providing assessment and guidance in improving projects or team building than more conventional open discussion at meetings, combined with periodic teleconferences and student presentations.

Experience within the CCMD indicated that something like the ambitious, integrative DARPA AIM program or other Foundational Engineering Problems [8,11-13] could not be effectively addressed in the context of the NSF funding level for an I/UCRC consortia. It takes major focused investment. Nonetheless, the fundamental contributions to methods and tools within the CCMD served to substantially advance ICME capabilities, as did the education of the next generation of graduate students and postdocs working in the context of the ICME paradigm. We see the need for companies to place higher priority on current and future workforce development relative to an emphasis on software tools. We also see the need to hire students educated and trained with the ICME 'mindset' and/or to engage universities in extended working relationships and student exchanges. Several CCMD stakeholders pursued such opportunities with excellent results, particularly those who engaged students as interns and/or committed engineering personnel to collaboration with the CCMD research programs to transfer technology and implement codes and methods in their organizations.

It became very clear to industry and government participants, as well as faculty and students involved in the CCMD, that ICME involves a change of culture and is not just an algorithmic addition to existing organizational processes and methods. Implementation of ICME within an organization requires buy-in and investment and a change of operations to incorporate modeling and simulation in a systematic way to accelerate decision-making in materials development. Moreover, materials development must integrate with manufacturing, quality control and automation, verification and validation, materials synthesis, processing, characterization, and property measurement. A computational materials design research center is therefore inherently limited in its breadth in addressing ICME. Nonetheless, bridging the fundamental gaps identified in the CCMD vision shown in Figure 6 is critical to facilitating the role of computation.

Along these same lines, the overarching goals of the CCMD align closely with those of the MGI announced by the White House in June 2011 [14]. Additionally, the fundamental concept underlying CCMD research reflects the essence of the opening statement 'A genome is a set of information encoded in the language of DNA that serves as a blueprint for an organism's growth and development. The word genome, when applied in non-biological contexts, connotes a fundamental building block towards a larger purpose'; the fundamental building block of materials is the individual phase. The CCMD has focused on developing approaches to model the properties and responses of individual phases as a function of process variables and to simulate their contributions to the properties and responses of microstructures consisting of polycrystalline/polyphase aggregates and their interfaces.

Conclusions

The vision of ICME is compelling in terms of value-added technology that reduces time to market for new products that exploit advanced, tailored materials. This case study considers the foundation, operation, and contributions of the joint Penn State-Georgia Tech CCMD, a NSF I/UCRC funded from 2005 to 2013. In spite of limitations on per project funding and constraints on mounting systematic, large-scale integrated materials design and development for specific materials systems, several key aspects of the current ICME and MGI emphases were established by the CCMD in this earlier time frame:

- Systematic development of the ICME workforce of the future;
- Building a culture of materials design and development, with increasing emphasis on computation; and
- New advances in computational tools to deal with diffusion, microstructure evolution, fracture and extreme value fatigue problems.

In advancing methodologies and tools to address ICME, it is essential to maintain a consistent long-term vision among industry, academia, and government. The required change of culture in academia towards materials research and development, as well as evolution of the curriculum, is an essential aspect with a relatively long time scale (perhaps a decade or more). Moreover, there is a critical need for industry to embrace this culture shift as well, which can be greatly facilitated by hiring students exposed to research initiatives such as the CCMD. Such students will employ modeling and simulation in industry practice. Furthermore, although engagement of materials suppliers within the CCMD was rather limited compared to involvement of OEMs, we view the materials supply chain as critical to the future of increasing the pace of materials discovery and development to meet ICME and MGI objectives. The materials supply chain might very well establish strong entrepreneurial leadership in the future of ICME, which will be advanced through innovative new business models and incentives.

A closing comment pertains to the broad multidisciplinary nature of ICME and materials innovation. Too often the discussions seem to revolve around the discipline of Materials Science & Engineering in the context of computational materials science, particularly in professional society and government planning venues. The materials innovation ecosystem is much broader. For example, the CCMD embraced from its inception the coupling of computational materials science, computational solid mechanics, and systems-based engineering design. These are rather disparate fields that typically do not strongly overlap in professional societies and archival journals. Students involved in the CCMD were witness to broad discussions across these disciplines to appreciate how scales and organizations can be bridged to achieve ICME goals by incorporating essential contributing elements of these different perspectives.

Following the CCMD vision in Figure 6, Penn State has engaged in expanding the concept of MaterialsGenome® and developing infrastructure for phase-based property data [30,31]. It is articulated that new data repository infrastructure is necessary so that when new models are developed or new experimental and computational information becomes available, the hierarchically structured materials property databases can be re-assessed efficiently to develop the new multicomponent descriptions. The ongoing effort of ESPEI (Extensible Self-optimizing Phase Equilibrium Infrastructure) serves as one example [32].

Georgia Tech has followed up on lessons learned with the CCMD experience in ICME to invest in developing a materials innovation infrastructure that broadly addresses the ICME vision and that of the MGI [14] as a highly multidisciplinary enterprise, with various expanded elements shown in Figure 9. The Institute for Materials (IMat) was founded in 2012 at Georgia Tech by CCMD Co-Director Dave McDowell (www.materials.gatech.edu). Reporting to the office of the Executive Vice President for Research, IMat is framing Georgia Tech's materials innovation ecosystem involving over 200 faculty members engaged in materials research to provide an institutional framework for collaboration in materials research. Connecting expertise, infrastructure, and resources that underpin the science and engineering of materials, the Institute serves as a hub for materials education and research across Georgia Tech and within the broader materials community. In addition to coordination of access and utilization of shared facilities within Georgia Tech, IMat is building a model for materials innovation that:

- Pursues a 'Materials + X' strategy in forming approaches that address grand challenges, with materials as an enabler of advances in energy, mobility, security, health, etc.;
- Promotes development of novel approaches to materials data sciences and informatics as part of a materials information infrastructure; and
- Fosters collaborative concepts for accelerating materials discovery, design, and development via high throughput computational and experimental strategies.

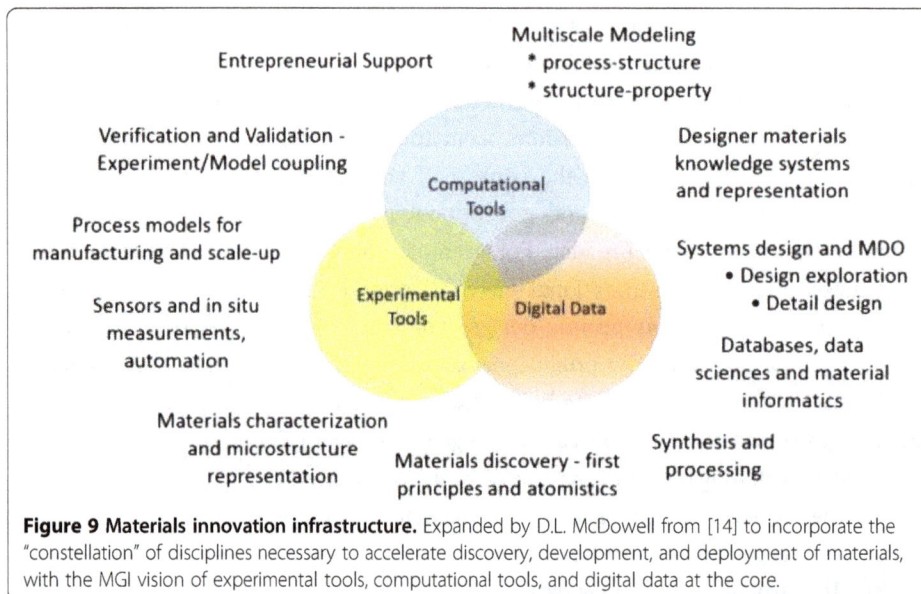

Figure 9 Materials innovation infrastructure. Expanded by D.L. McDowell from [14] to incorporate the "constellation" of disciplines necessary to accelerate discovery, development, and deployment of materials, with the MGI vision of experimental tools, computational tools, and digital data at the core.

In addition to hosting a 28 March 2014 Southeastern U.S. regional workshop for the MGI, IMat collaborated with the University of Wisconsin-Madison and the University of Michigan to co-organize the workshop 'Building an Integrated MGI Accelerator Network', held at Georgia Tech 5 to 6 June 2014. An initiative that arose through discussions with the White House Office of Science and Technology Policy, the purpose of the Accelerator Network is to launch a nationwide dialogue to connect centers, institutes, and future efforts to fulfill the MGI (http://acceleratornetwork.org/).

Competing interests

The authors declare that they have no competing interests.

Authors' contributions

ZKL and DLM co-directed the CCMD with NSF support from 2005 to 2013, and both contributed substantially to this case study. DLM drafted this manuscript with significant additional input from ZKL, with a number of iterations between them to reach final form. Both authors read and approved the final manuscript.

Acknowledgements

The authors are grateful for the long-term support of the NSF Industry/University Cooperative Research Center for Computational Materials Design (CCMD), including dues contributions of CCMD members, through grants IIP-0433033 (Penn State), IIP-0541674 (Penn State) and IIP-541678 (Georgia Tech) from 2005–2010, and IIP-1034965 (Penn State) and IIP-1034968 (Georgia Tech) from 2010 to 2013. The authors would like to thank our collaborators at Penn State and Georgia Tech for their enthusiastic and innovative contributions to CCMD projects and meetings over the years, including Co-PIs of the CCMD planning, phase I, and phase II proposals (Long-Qing Chen, Qiang Du, James Kubicki, Evangelos Manias, Padma Raghavan, and Jorge Sofo at Penn State, and Hamid Garmestani, Farrokh Mistree, Richard Neu, and Min Zhou at Georgia Tech), various additional Penn State and Georgia Tech faculty involved in CCMD proposals and projects, and students who conducted research with CCMD support. ZKL also acknowledges Penn State for further reduced overhead rate and Penn State MRI (Materials Research Institute) that provided partial support of the CCMD administrative staff. DLM also acknowledges the support of the Carter N. Paden, Jr. Distinguished Chair in Metals Processing. Any opinions, findings, and conclusions or recommendations expressed in this publication are those of the authors and do not necessarily reflect the views of the NSF, Penn State, or Georgia Tech.

Author details

[1]Department of Materials Science and Engineering, The Pennsylvania State University, University Park, Pennsylvania, PA 16802, USA. [2]Woodruff School of Mechanical Engineering, School of Materials Science and Engineering, Georgia Institute of Technology, Atlanta, GA 30332-0405, USA.

References

1. Pollock TM, Allison JE, Backman DG, Boyce MC, Gersh M, Holm EA, LeSar R, Long M, Powell AC IV, Schirra JJ, Whitis DD, Woodward C (2008) Integrated computational materials engineering: a transformational discipline for improved competitiveness and national security. National Materials Advisory Board, NAE, National Academies Press, Washington, DC, ISBN-10: 0-309-11999-5
2. Olson GB (1997) Computational design of hierarchically structured materials. Science 277:1237–1242
3. McDowell DL, Story TL (1998) New directions in materials design science and engineering (MDS&E). Report of a NSF DMR-sponsored workshop held at Georgia Tech, October 19-21, http://www.me.gatech.edu/paden/material-design/md_se.pdf, accessed December 2, 2014
4. Liu ZK, Chen LQ, Spear KE, Pollard C (2003) An integrated education program on computational thermodynamics, kinetics, and materials design, an article from the Dec. 2003 JOM-e, a Web-Only Supplement to JOM, TMS. http://www.tms.org/pubs/journals/JOM/0312/Liull/Liull-0312.html. Accessed December 2, 2014
5. Liu ZK, Chen LQ, Raghavan P, Du Q, Sofo JO, Langer SA, Wolverton C (2004) An integrated framework for multi-scale materials simulation and design. J Comput Aided Mater Des 11(2–3):183–199
6. Liu ZK (2009) First principles calculations and Calphad modeling of thermodynamics. J Phase Equilib Diffus 30:517–534
7. Liu ZK (2009) A materials research paradigm driven by computation. JOM 61(10):18–20
8. McDowell DL (2007) Simulation-assisted materials design for the concurrent design of materials and products. JOM 59:21–25
9. McDowell DL, Olson GB (2008) Concurrent design of hierarchical materials and structures. Sci Model Simul 15:207–240
10. McDowell DL, Panchal JH, Choi HJ, Seepersad CC, Allen JK, Mistree F (2009) Integrated design of multiscale, multifunctional materials and products, 1st edn. Elsevier, Oxford, p 392. ISBN 978-1-85617-662-0
11. Apelian D, Alleyne A, Handwerker CA, Hopkins D, Isaacs JA, Olson GB, Vidyanathan R, Wolf SD (2004) Accelerating technology transition: bridging the valley of death for materials and processes in defense systems. National Materials Advisory Board, NAE, National Academies Press, Washington, DC, ISBN-10: 0-309-09317-1
12. Allison J, Backman D, Christodoulou L (2006) Integrated computational materials engineering: a new paradigm for the global materials profession. JOM 58:25–27

13. McDowell DL, Backman D (2010) Simulation-assisted design and accelerated insertion of materials, Ch. 19. In: Ghosh S, Dimiduk D (eds) Computational methods for microstructure-property relationships. Springer, New York. ISBN 978-1-4419-0642-7

14. (2011) The materials genome initiative for global competitiveness, office of science and technology policy. National Science and Technology Council, http://www.whitehouse.gov/sites/default/files/microsites/ostp/materials_genome_initiative-final.pdf, accessed December 2, 2014

15. Liu ZK, McDowell DL (2006) Center for computational materials design (CCMD) and its education vision. TMS MS&T, Cincinnati, OH

16. McDowell DL (2006) Simulation and robust design of materials. TMS MS&T, Cincinnati, OH

17. McDowell DL, Mistree F, Allen JK (2007) Prospects and challenges for materials design. Mechanics and materials modeling and materials design methodologies, symposium in honor of Dr. Craig Hartley's 40 years of contributions to the field of mechanics and materials science. TMS Annual Meeting & Exhibition, Orlando, FL

18. Liu ZK (2007) Integrating forward simulation and inverse design of materials. TMS webcast, http://iweb.tms.org/forum/messageview.aspx?catid=97&threadid=1094&enterthread=y

19. Liu ZK (2007) Properties of individual phases by first-principles calculations and CALPHAD modeling. Eastern New York ASM/TMS Annual Symposium. Computational Materials Design, GE Global Research Center, Niskayuna, NY

20. McDowell DL (2008) Multiscale modeling and materials design. TMS 2008 9th Global Innovations Symposium on Trends in ICME, New Orleans, LA

21. McDowell DL (2008) Multiscale modeling in multilevel materials design. Kickoff lecture, symposium on computational materials design via multiscale modeling. Session on New Approaches Toward Multiscale Materials Design, MRS Fall Meeting, Boston, MA

22. McDowell DL (2009) Some comments on materials design education. MS&T, Pittsburgh, OH

23. Liu ZK, McDowell DL (2009) Materials research paradigm driven by computation. MS&T, Pittsburgh, OH

24. McDowell DL (2010) Robust materials design and multiscale simulation: distinct but complementary pursuits. Tools, models, databases and simulation tools developed and needed to realize the vision of ICME: material model and simulation tools, part II. MS&T, Houston, TX

25. McDowell DL (2011) Critical path issues in ICME. Models, databases, and simulation tools needed for the realization of integrated computational materials engineering, Proc. Symposium held at MS&T 2010, Houston, Tx, S.M. Arnold and T.T. Wong, eds., ASM International, 31-37

26. Liu ZK (2012) Materials genome: building blocks of materials. TMS Annual Meeting, Orlando, FL

27. McDowell DL (2012) Simulation-based strategies to support alloy design for fatigue resistance. Symposium on Integrative Materials Design: Performance and Sustainability. TMS Annual Meeting, Orlando, FL

28. McDowell DL (2013) Modeling inelastic behavior of metals at multiple scales to support materials design. MS&T '13, Montreal, Quebec, Canada

29. Panchal JH, Kalidindi SR, McDowell DL (2013) Key computational modeling issues in ICME. Comput Aided Des 45(1):4–25

30. Liu ZK (2014) Perspective on Materials Genome®. Chin Sci Bull 59(15):1619–1623

31. Campbell EC, Kattner RU, Liu ZK (2014) The development of phase-based property data using the CALPHAD method and infrastructure needs. Integrating Materials and Manufacturing Innovation 3:12. doi:10.1186/2193-9772-3-12

32. Shang SL, Wang Y, Liu ZK (2010) ESPEI: extensible, self-optimizing phase equilibrium infrastructure for magnesium alloys. In: Agnew SR, Neelameggham NR, Nyberg EA, Sillekens WH (eds) Magnesium technology 2010, Seattle. WA, Minerals, Metals and Materials Society/AIME, 184 Thorn Hill Road, Warrendale, PA, pp 617–622

7

Effect of the curing process on the transverse tensile strength of fiber-reinforced polymer matrix lamina using micromechanics computations

Royan J D'Mello[1,2], Marianna Maiarù[1,2] and Anthony M Waas[1,2]*

*Correspondence:
awaas@aa.washington.edu
[1] Composite Structures Laboratory, Department of Aerospace Engineering, University of Michigan, 1320 Beal Avenue, Ann Arbor, MI 48109-2140, USA
[2] William E. Boeing Department of Aeronautics and Astronautics, University of Washington, Seattle, WA 98195-2400, USA

Abstract

The effect of the curing process on the mechanical response of fiber-reinforced polymer matrix composites is studied using a computational model. Computations are performed using the finite element (FE) method at the microscale where representative volume elements (RVEs) are analyzed with periodic boundary conditions (PBCs). The commercially available finite element (FE) package ABAQUS is used as the solver, supplemented by user-written subroutines. The transition from a continuum to damage/failure is effected by using the Bažant-Oh crack band model, which preserves mesh objectivity. Results are presented for a hexagonally packed RVE whose matrix portion is first subjected to curing and subsequently to mechanical loading. The effect of the fiber packing randomness on the microstructure is analyzed by considering multi-fiber RVEs where fiber volume fraction is held constant but with random packing of fibers. The possibility of failure is accommodated throughout the analysis—failure can take place during the curing process prior to the application of in-service mechanical loads. The analysis shows the differences in both the cured RVE strength and stiffness, when cure-induced damage has and has not been taken into account.

Keywords: Curing; Stress evolution; Periodic boundary condition; Crack band model

Background

Fiber-reinforced polymer matrix composites (FRPCs) are high-strength and lightweight advanced materials widely used in the aerospace and automotive industries. Since FRPCs are manufactured by curing the matrix that surrounds the interspersed fibers, good understanding of the matrix state during the curing process is necessary to have sufficient control over the quality of the cured product. The mechanical properties of the matrix during curing can be altered by the presence of fibers and also by details of the curing cycle. The curing matrix undergoes shrinkage due to chemical processes, which gives rise to self-equilibrating internal stresses. Plepys and Farris [1] and Plepys et al. [2] have used finite element calculations using incremental elasticity to show tensile residual stress buildup of up to 28 MPa post cure in a three-dimensionally constrained Epon 828 epoxy resin. Merzlyakov et al. [3] reported the development of tensile stresses in a constrained thermosetting resin system undergoing cure and also quantified the variation of these tensile stresses during subsequent thermal cycling. Depending on the constituent chemistry of the matrix, the thermal cycle prescribed, and the fracture and strength properties of a curing matrix, a fiber-reinforced composite can and may undergo damage and cracking in the matrix during the cure cycle. Chekanov et al. [4] have reported various types of defects that may form in a constrained epoxy resin system undergoing curing. Rabearison et al. [5] studied the curing of a thick epoxy tube using a finite element model and concluded that high stress gradients developed during differential curing can cause cracking. Therefore, the state of the matrix within a cured FRPC structure exhibits *in situ* matrix properties, which are effective properties of the matrix that take into account imperfections caused in the matrix due to the cure process, including the presence of residual stresses. That is, the *in situ* matrix properties, where the matrix is treated as a 'new' material with a reference configuration that corresponds to the post-cured state, deviate from idealized or 'virgin' matrix properties of the bulk matrix. The *in situ* matrix properties can be extracted from an inverse analysis [6] through the uniaxial tensile response of a $\pm 45°$ laminate, and this is convenient in engineering analysis of cured composites. Song and Waas [7] have shown that the use of bulk matrix properties in numerical predictions of compression response of a 2D triaxially braided composite RVE can lead to erroneous results - the computed compressive strength being noticeably higher than the experimentally measured strength. They observed that the tow kinking failure mode, which controls the compression strength was found to be sensitive to the nonlinear shear response of the matrix. Cure shrinkage in the matrix surrounded by randomly dispersed fibers can also influence the final shape of the structure [8]. Therefore, it is necessary to have good knowledge of the influence of the cure cycle on the subsequent mechanical response of the laminate. For a particular fiber-matrix laminate system, the optimal cure cycle can be identified such that the cured product has the highest strength and stiffness. Efforts to optimize various aspects of the cure cycle for mitigating the residual stresses generated during cure can be found in the studies of Li et al. [9], Gopal et al. [10], and White and Hahn [11].

In the present investigation, the effects of the cure cycle on possible damage accumulation during cure and subsequent in-service performance at the microstructural level are studied. A hexagonally packed representative volume element (RVE) having a total of two fibers (one full center fiber and quarter fibers at four corners) with different volume fractions, and a randomly packed RVE having multiple fibers are studied. First, the influence of fiber volume fraction on the strength of the cured RVE using the hexagonally packed RVE with two fibers is studied. Next, the effect of the randomness of the packing for RVEs having fixed volume fraction is investigated. For illustrating the findings of this study, the strength investigated is the transverse tensile strength (S_{22}^+), which is obtained by mechanically loading each of the virtually cured RVEs along the transverse direction under tension. Then, the initial slope and peak stress value of the nominal stress-strain response are the transverse stiffness E_{22} and transverse tensile strength S_{22}^+, respectively. For low to moderate fiber volume fractions, the transverse stiffness is controlled by matrix stiffness (see [12]). The transverse tensile strength associated with *transverse matrix cracking* is controlled by a combination of factors such as matrix tensile strength, matrix fracture toughness, fiber packing, and adhesion strength between fibers and the matrix. Hence, it is expected that both E_{22} and S_{22}^+ are influenced by the details of the cure process.

Methods

Cure process

The curing process of a thermoset polymer can be divided into two parts: The first part consists of the chemical reaction, heat generation, and conduction. The second is the generation of self-equilibrating stresses and development of the structural integrity via the evolution of matrix stiffness. The stress generation has been modeled by Mei [13], Mei et al. [14], and Heinrich et al. [15]. The degree of cure (ϕ) of the matrix is defined as $\phi = H(t)/H_r$, where $H(t)$ is the heat generated up to time t, and H_r is the total heat of reaction at the end of the cure cycle. Mathematically, the rate of cure $\left(\dfrac{d\phi}{dt}\right)$ can be expressed as,

$$\frac{d\phi}{dt} = f(T, \phi) \tag{1}$$

where $f(T, \phi) \geq 0$ is a function. The evolution of temperature (T) and degree of cure (ϕ) for the matrix material system is determined through a coupled system that considers the heat equation and an empirical curing law or can be supplied from the output of a simulation that takes into account a cure kinetics model. Kamal [16] has proposed a semi-empirical expression for the function $f(T, \phi)$ in terms of Arrhenius terms that depend on temperature

$$f(T, \phi) = \left[A_1 \exp\left(\frac{\Delta E_1}{TR}\right) + A_2 \exp\left(\frac{\Delta E_2}{TR}\right) \phi^m \right] (1 - \phi)^n \tag{2}$$

where T is temperature, R is the gas constant, and ΔE_1 and ΔE_2 are activation energies. The frequency-like constants A_1, A_2 and exponents m and n, in theory, have to be determined by fitting the above equation to the experimental data. However, due to the complexity of the function $f(T, \phi)$, a general closed formed solution to Equation 1 is elusive, and often times, this differential equation has to be solved using some numerical method. Assuming the form for $f(T, \phi)$ in Equation 2 by setting $m = A_2 = \Delta E_2 = 0, n = 1$ and under isothermal conditions, an explicit relation between the degree of cure and time can be found as a solution to the differential equation 1, which is

$$\phi(t) = 1 - \exp(-\lambda t) \tag{3}$$

where the Arrhenius parameter $\lambda = A_1 \exp\left(\dfrac{-\Delta E_1}{TR}\right)$. Cure data as a function of time for Epon 862/Epikure 9553 resin under isothermal conditions are chosen for the present work and are available in [15]. The constants obtained by curve fitting with experimental data at various temperatures are as follows: $A_1 = 3.62 \times 10^{11}$ s^{-1} and $\Delta E_1 = 8.854 \times 10^4$ J.

During curing, the matrix heats up due to an exothermic chemical reaction and due to conduction from the heating source at the boundary. This process can be modeled using the equation

$$\rho c \frac{\partial T}{\partial t} = \frac{\partial}{\partial x_i} \left(\kappa(T, \phi) \frac{\partial T}{\partial x_i} \right) + \rho H_r \frac{\partial \phi}{\partial t} \tag{4}$$

where ρ is the mass density, c_p is the specific heat, and κ is the thermal conductivity. The evolution of self-equilibrating stresses $\sigma_{ij}(t)$ during curing is included in the analysis by using a model proposed by Heinrich et al. [15]:

$$\underline{\underline{\sigma}}(t) = \int_0^t \frac{d\phi}{ds} \underline{\underline{1}} \Big[K(s) tr \left(\underline{\underline{\varepsilon}}(t) - \underline{\underline{\varepsilon}}(s) + \underline{\underline{\varepsilon_c}}(s) - \underline{1}\alpha(s)\Delta T(t,s) \right)$$

$$+ 2\mu(s) \left(\underline{\underline{\varepsilon}}(t) - \underline{\underline{\varepsilon}}(s) + \underline{\underline{\varepsilon_c}}(s) - \underline{\underline{1}}\frac{1}{3} tr\{\underline{\underline{\varepsilon}}(t) - \underline{\underline{\varepsilon}}(s) + \underline{\underline{\varepsilon_c}}(s)\} \right) \Big] ds \qquad (5)$$

$$+ (1 - \phi(t))K(0)tr(\underline{\underline{\varepsilon}}(t) - \underline{1}\alpha(0)\Delta T(t))\underline{1}$$

where K, μ, α, and ε_c are the per-network bulk modulus, shear modulus, coefficient of thermal expansion, and cure shrinkage, respectively. The first term having the integral is the contribution to stress evolution due to the curing matrix, whereas the second term captures the contribution of the uncured liquid resin. The constants $K(0)$ and $\alpha(0)$ correspond to the bulk modulus and coefficient of thermal expansion of the liquid resin, respectively. The coefficient of thermal expansion $\alpha(\phi)$ of the curing matrix is assumed to have a constant value of 61×10^{-6} m/mK. As shown by Heinrich et al. [15], the per-network properties can be obtained from experimentally measured values of the plane wave modulus (M_{exp}) and shear modulus (μ_{exp}) for the curing matrix as

$$M(\phi) = \frac{dM_{exp}}{d\phi} + K_{exp}(0)$$

$$\mu(\phi) = \frac{d\mu_{exp}}{d\phi} \qquad (6)$$

The moduli values M_{exp} and μ_{exp} are measured as a function of time by concurrent Raman and Brillouin light scattering for the pure resin, that is, for a resin curing in the absence of fibers. These moduli are assumed to correspond to the virgin matrix as a function of degree of cure. The effect of the presence of fibers around the matrix on matrix degradation during cure will be demonstrated later in this paper. Once $M(\phi)$ and $\mu(\phi)$ are known, the per-network bulk modulus $K(\phi)$ can be obtained from the isotropic material relation $K = M - \frac{4}{3}\mu$. The per-network shrinkage strain $\varepsilon_c(\Phi)$ up to a certain degree of cure $\phi = \Phi$ is given by

$$\varepsilon_c = \frac{1}{3K(\Phi)} \left[\left(\varepsilon(\Phi) - (1 - \Phi)\frac{d\varepsilon(\Phi)}{d\Phi} \right) K_{exp} - \frac{d\varepsilon(\Phi)}{d\Phi} \int_0^\Phi M(\phi)d\phi \right] \qquad (7)$$

A gravimetric test method (see [17]) can be used to obtain shrinkage of all networks $\varepsilon(\Phi)$. A 2% per-network cure shrinkage has been chosen for the present investigation.

Damage during cure

During curing, the matrix gradually solidifies (stiffness increases) and simultaneously contracts (cure shrinkage) due to network formation. Residual stresses develop in the matrix owing to cure shrinkage and thermal strains. Depending on the magnitude of tensile stresses developed, the degree of cure (ϕ), and the rate of cure $\left(\frac{d\phi}{dt} \right)$, the material may crack locally during curing. A crack band model is used to simulate the possibility of tensile cracking during the curing of the matrix. The critical tensile stress for cracking typically increases with the degree of cure. If certain matrix regions crack locally, it would result in a reduction in the matrix stiffness in that local region along with some energy dissipated due to cracking. Such a reduction in local matrix stiffness can control the mechanical properties of the cured RVE. Two assumptions are enforced because the degree of cure and the coefficient of thermal expansion of a partially cured local volume of material with microcracks are unknown and physically this local volume does not represent a continuum in the strictest sense. First, if a certain local volume of material cracks, it is assumed that no further curing can take place in that local volume. Second, it is assumed that if cracking occurs locally, the local cracked volume cannot expand or contract under temperature variations. In the context of the finite element framework that is used to numerically simulate cure-induced damage, the local volume is a single finite element. Therefore and because the crack band method is used, mesh objectivity is included in the formulation.

At the end of step II, the curing process is complete. In step III, the cured RVE (containing cracks or not, as the case maybe) is subjected to transverse tension loading along the 2-direction. The objective here is to compute the strength (S_{22}^+) and stiffness (E_{22}) of the virtually cured RVE. Based on the temperature and cure parameters, computation of the stress evolution during cure (step II) and strength calculation based on mechanical loading (step III) is done in a unified step in the commercial software ABAQUS/Standard [18]. In this study, it is assumed that cracking in the curing matrix can occur only for $\phi > 0.2$ and only under tensile stresses.

The crack band model of Bažant and Oh [19] is used to model failure in the matrix. This model assumes that once the critical fracture stress σ_{cr} has been reached, microcracks are formed and the additional opening due to cracking is smeared over a band of material. Here the width of that band is taken to be that which lies within an element and perpendicualr to the crack plane. The maximum principal stress criterion is used to determine the failure initiation. In the post-peak regime, the traction-separation law controls the behavior of the damaging material as shown in Figure 1 and the stiffness of the material (matrix) is reduced using the secant value. In the present investigation, σ_{cr} is assumed to be independent of ϕ. However, in reality, it is expected that the strength would vary with ϕ. Under mode I cracking, the energy dissipated during the fracturing process is the critical mode I energy release rate (G_{IC}) given by

$$G_{IC} = \int_0^{\delta_f} \sigma_{11}(\delta)d\delta = h \int_0^{\varepsilon_f} \sigma_{11}(\varepsilon_{11})d\varepsilon \qquad (8)$$

where stress σ_{11} and ε_{11} are the maximum principal stress and strain values, respectively, and the maximum separation $\delta_f = h\varepsilon_f$ where ε_f corresponds to the critical failure strain of the material (accompanied by complete loss of stiffness). Here, h is the characteristic element length that preserves *mesh objectivity* (see [20]), defined by prescribing a normalized value of G_{IC} for each element such that $g_{IC} = \dfrac{G_{IC}}{h}$. Consequently, the value of g_{IC}

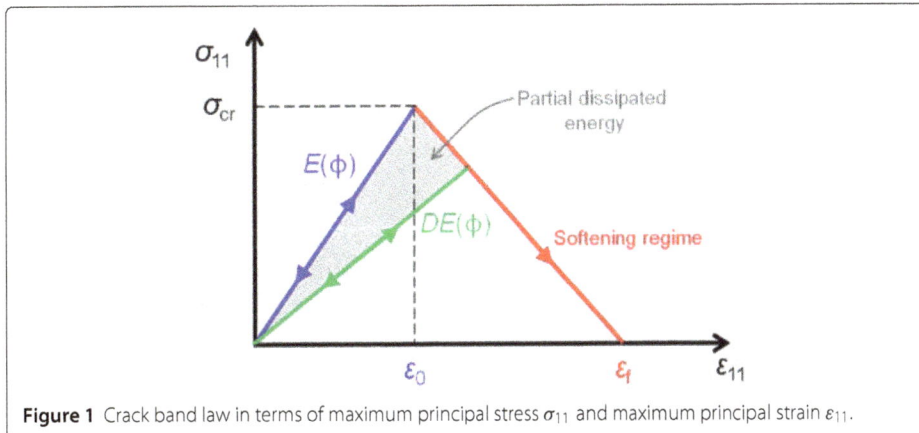

Figure 1 Crack band law in terms of maximum principal stress σ_{11} and maximum principal strain ε_{11}.

equals the area under the $\sigma_{11} - \varepsilon_{11}$ law shown in Figure 1. The value of G_{IC} is chosen to be 0.6 N/mm in all the computations. For a given epoxy system, the values of G_{IC} and σ_{cr} have to be obtained from an experiment, each as a function of the degree of cure ϕ.

From the crack band model formulation, the stiffness reduction factor D with $(0 \leq D \leq 1)$ for a material with initial stiffness $E = E(\phi)$ which is now in the softening region of the traction-separation law is computed as

$$D = \frac{\sigma_{cr}}{E(\varepsilon_f - \varepsilon_{cr})} \left(\frac{\varepsilon_f}{\varepsilon_{11}} - 1 \right) \tag{9}$$

where ε_{11} is the current maximum principal strain value. Thus, $D = 1$ corresponds to no damage, $0 < D < 1$ corresponds to damage but no two-piece failure, while $D = 0$ would indicate complete failure. This D parameter will be used to quantify the extent of stiffness reduction after cure has been completed (i.e., at the end of step II).

Boundary conditions

During curing and mechanical loading, the RVE is subjected to periodic boundary conditions, in concert with the assumption that the RVE is a small volume within an infinite medium. The use of periodic boundary conditions for fiber-reinforced RVEs can be found in the studies of Gonzalez and Llorca [21] and Xia et al. [22], among others. During the cure process (step II), the RVE boundaries are allowed to contract or expand. The RVE can contract or expand depending on temperature change and can contract due to cure shrinkage.

Consider an arbitrary cuboid RVE in the undeformed configuration having lengths L_1, L_2, and L_3 along the x_1, x_2, and x_3 directions with one corner point placed at the origin $(0, 0, 0)$. Then, the equations corresponding to the 3D periodic boundary conditions are

$$\begin{aligned}
u_1(L_1, x_2, x_3) - u_1(0, x_2, x_3) &= \epsilon_{11} L_1 \\
u_2(L_1, x_2, x_3) - u_2(0, x_2, x_3) &= 2\epsilon_{12} L_1 \\
u_3(L_1, x_2, x_3) - u_3(0, x_2, x_3) &= 2\epsilon_{13} L_1 \\
u_1(x_1, L_2, x_3) - u_1(x_1, 0, x_3) &= 2\epsilon_{21} L_2 \\
u_2(x_1, L_2, x_3) - u_2(x_1, 0, x_3) &= \epsilon_{22} L_2 \\
u_3(x_1, L_2, x_3) - u_3(x_1, 0, x_3) &= 2\epsilon_{23} L_2 \\
u_1(x_1, x_2, L_3) - u_1(x_1, x_2, 0) &= 2\epsilon_{31} L_3 \\
u_2(x_1, x_2, L_3) - u_2(x_1, x_2, 0) &= 2\epsilon_{32} L_3 \\
u_3(x_1, x_2, L_3) - u_3(x_1, x_2, 0) &= \epsilon_{33} L_3
\end{aligned} \tag{10}$$

u_1, u_2, and u_3 are the displacements of the RVE boundary along the x_1, x_2, and x_3 directions, respectively, and ϵ_{ij} are the tensorial strains.

Analysis procedure

In summary, the analysis procedure is divided into three steps as shown in Figure 2.

1. *Step I*: A thermochemical analysis is performed using the cure parameters described earlier. Temperature cycle, the degree of cure, and the cure rate in the matrix are provided. Since the RVE dimensions are on the micron scale, there is little to no variation in the temperature field across the RVE. The temperature

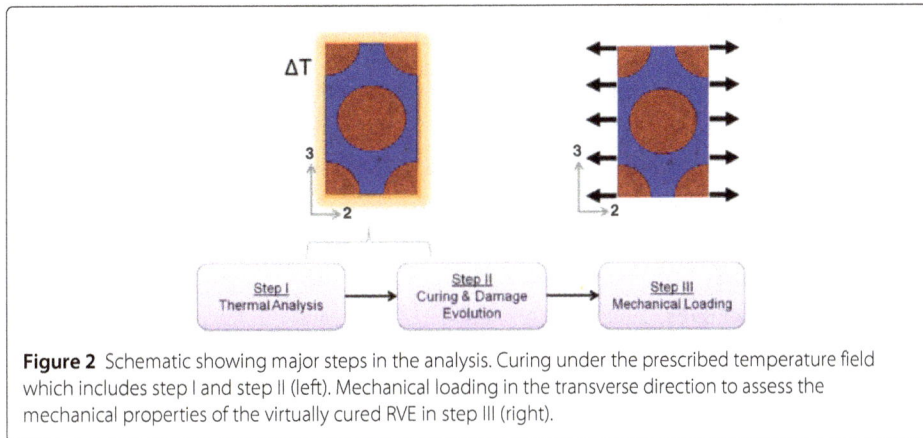

Figure 2 Schematic showing major steps in the analysis. Curing under the prescribed temperature field which includes step I and step II (left). Mechanical loading in the transverse direction to assess the mechanical properties of the virtually cured RVE in step III (right).

profile, the degree of cure (ϕ), and the rate of cure $\left(\dfrac{d\phi}{dt}\right)$ used in the present study are shown in Figure 3.

2. *Step II*: The stress evolution calculations are preformed as described in Equation 5. Shrinkage during cure is modeled using Equation 7. At the end of this step, we have a virtually cured solid. Possibility of damage during curing is taken into account

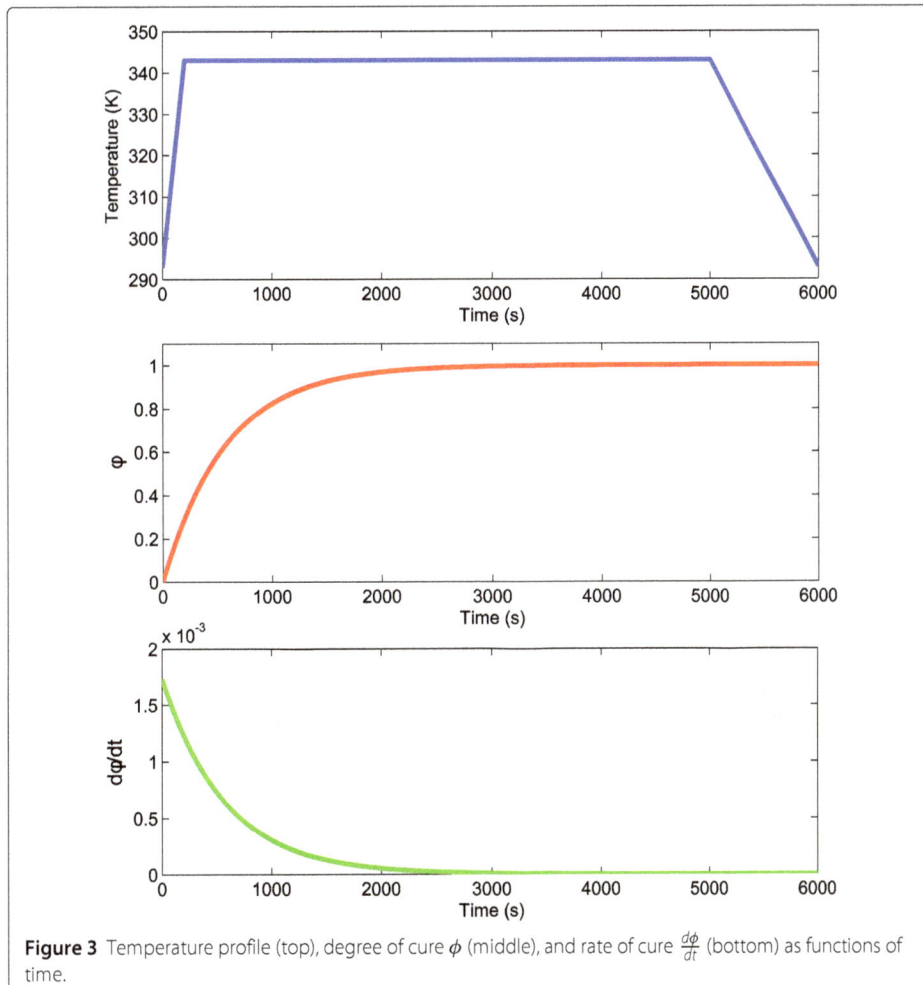

Figure 3 Temperature profile (top), degree of cure ϕ (middle), and rate of cure $\frac{d\phi}{dt}$ (bottom) as functions of time.

using a crack band model. Periodic boundary conditions are enforced throughout this step.

3. *Step III*: The virtually cured RVE is subjected to transverse tensile loading (with periodic boundary conditions in place) to back out the stiffness and strength. Again, the crack band model is used to simulate tensile failure, and periodic boundary conditions are enforced during this step.

Results and discussion

Hexagonally packed fiber RVEs

Three 3D hexagonally packed RVEs with fiber volume fractions (V_f) of 0.5, 0.6, and 0.7 are studied. These RVEs are first subjected to the curing cycle (steps I and II) and then to tensile loading (step III) in the transverse direction. The latter step leads to the determination of the transverse stiffness E_{22} and tensile strength S_{22}^+ of the virtually cured RVEs. The analysis is done using the finite element software ABAQUS/Standard. The stress evolution expression along with the crack band model is implemented using ABAQUS/Standard's user subroutine UMAT. In each of the RVEs shown in Figure 4, the thickness t along the fiber direction is chosen to be 0.30 μm and carbon fibers are 6 μm in diameter. Both the fiber and the matrix are modeled as isotropic solids. Young's modulus and Poisson's ratio of the fibers are taken to be 200 MPa and 0.3, respectively.

For each of the RVEs shown in Figure 4, three critical fracture strength values (σ_{cr}) of 20, 30, and 45 MPa are chosen which are independent of the degree of cure ϕ, while the critical mode I energy release rate G_{IC} is chosen to be 0.6 N/mm. The strength and toughness are assumed here to be independent of the degree of cure (ϕ). The objective of this portion of the study is to understand how the strength and stiffness of the cured product change with changes in fiber volume fraction and changes in the imposed critical fracture strength during cure. For a given RVE in step II, the matrix tensile stresses can exceed σ_{cr} and microcracks appear leading to a reduction in stiffness. To assess the amount of cure-induced damage, we can keep track of the stiffness reduction factor D at various times during the curing process. Figure 5 shows the average D value for each of the RVEs undergoing cure. Recall that $D = 1$ corresponds to no loss of instantaneous stiffness, whereas $D = 0$ corresponds to the complete loss of stiffness. Here, for each of the RVEs, the D values drop first for the case with $\sigma_{cr} = 20$ MPa followed by the case with $\sigma_{cr} = 30$ MPa and lastly by the case with $\sigma_{cr} = 45$ MPa. This is expected as damage would occur first in the RVE that has the lowest critical strength. Consequently, the RVEs

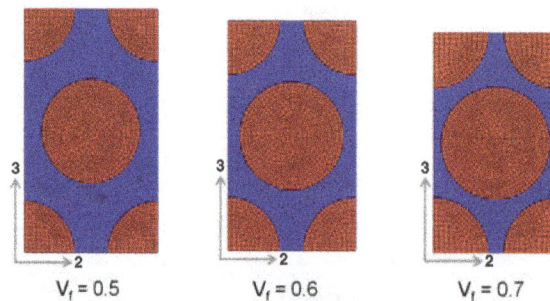

Figure 4 Hexagonally packed fiber-matrix RVEs with fiber volume fractions of 0.5, 0.6, and 0.7. Notice that the dimensions of the fiber are held fixed across these three RVEs.

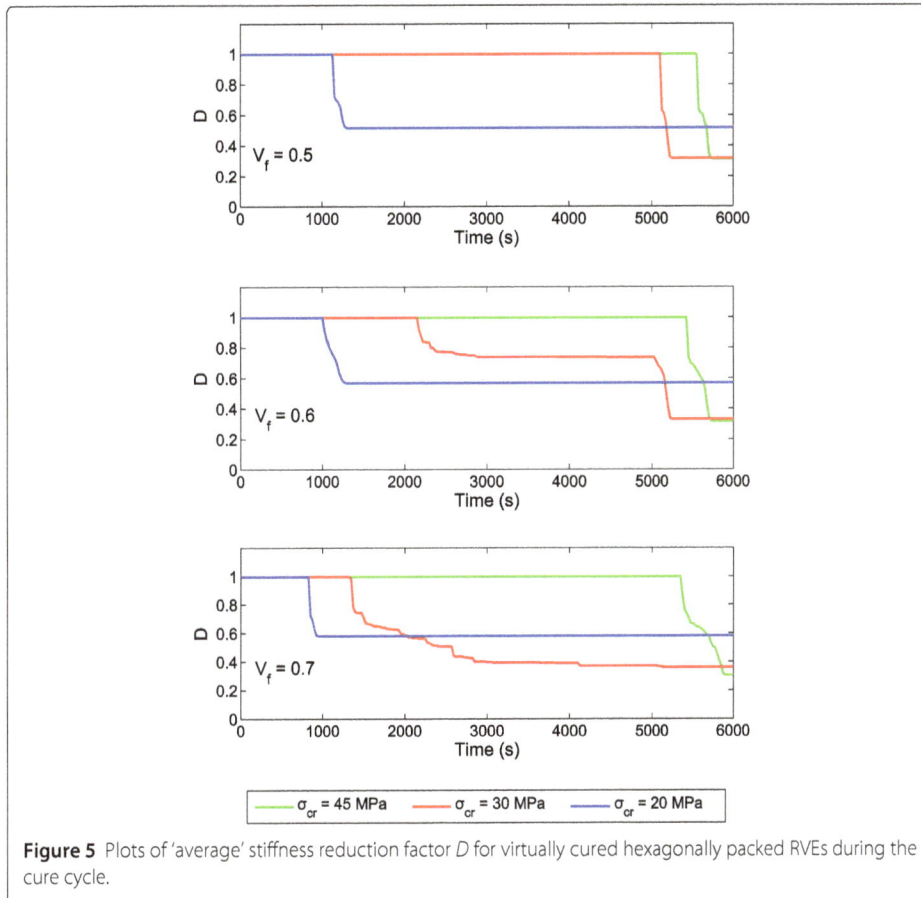

Figure 5 Plots of 'average' stiffness reduction factor D for virtually cured hexagonally packed RVEs during the cure cycle.

with $\sigma_{cr} = 20$ MPa are also the first to stop curing (locally, at those locations where cracking has occured), on account of microcrack formation. The simulations with σ_{cr} values of 30 and 45 MPa first exhibit microcracks during the cooling phase ($5,000$ s $\leq t \leq 6,000$ s) of the cure cycle where additional shrinkage occurs due to cooling. It is interesting to note that although microcracks appear last for the case with $\sigma_{cr} = 45$ MPa, the drop in D is more drastic when compared to drops corresponding to the other two cases. Hence, at the end of the cure cycle, for each RVE, the extent of damage varies inversely with the critical fracture strength σ_{cr} of the curing matrix. The spatial variation of D at the end of cure is shown in Figure 6. Even though the RVE is symmetric about a vertical and horizontal line passing through its center, there is nonhomogeneity in the contour of D across the matrix. The nonhomogeneity in D arises because the stress distribution in the RVE does not strictly follow the symmetry present in the hexagonal packing during cure on account of small numerical differences. Hence, once cracking starts at locations where stresses are highest, this breaks the symmetry in stress distribution, thus leading to subsequent nonhomogeneity in D as the curing progresses. In these curing simulations, two-piece failure (corresponding to $D = 0$) was not observed in the cured matrix.

The virtually cured RVEs are now loaded in tension along the transverse direction in step III. As in the previous step, periodic boundary conditions are enforced. The nominal stress-strain ($\sigma_{22} - \varepsilon_{22}$) response is shown in Figure 7. Each of the cured RVEs exhibits a fairly linear response during the initial stages of loading (up to nominal strain), followed

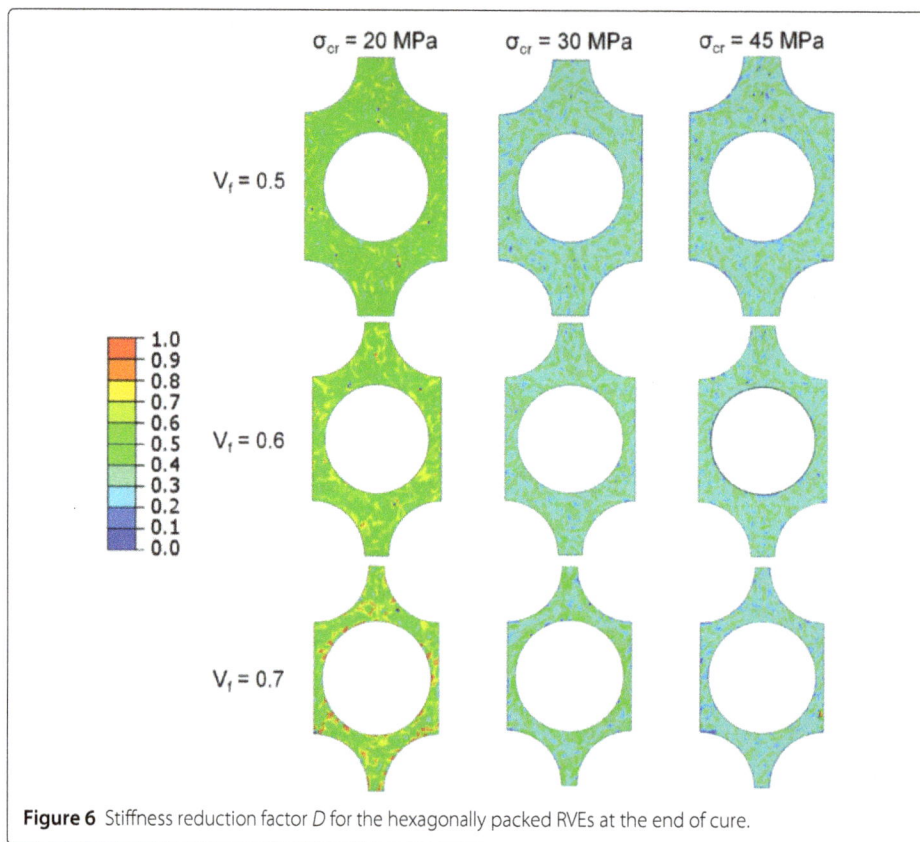

Figure 6 Stiffness reduction factor D for the hexagonally packed RVEs at the end of cure.

by a nonlinear softening response before attaining the peak. Past the peak, a rapid drop in stress is observed. The peak stress values correspond to the transverse strength S_{22}^+ of the virtually cured RVEs. In the case where cure-induced damage is ignored, and when the RVEs are loaded under tension in the transverse direction, the resulting stress-strain response is shown in Figure 8. These RVEs also exhibit a fairly linear response during the initial stages of loading. However, the extent of nonlinearity present before the peak is much lesser than the case when cure-induced damage is accounted for (see Figure 7). The RVEs with no cure-induced damage exhibit higher global stiffness compared to those when cure-induced damage is taken into account. This is as expected. In the post-peak region, the crack paths for simulations with $\sigma_{cr} = 20$ MPa for virtually cured RVEs and for RVEs where cure-induced damage has not been taken into account are shown in Figure 9 and in Figure 10, respectively. Figure 11 shows the variation of the initial stiffness (E_{22}) of RVEs under mechanical loading. For a given RVE with volume fraction held fixed, the lowest stiffness is exhibited by the RVE having the highest σ_{cr} value of 45 MPa in step II. Recall that from Figure 5, this case with $\sigma_{cr} = 45$ MPa had the lowest value of D at the end of the cure cycle. Thus, the stiffness reduction factor D is seen to have a positive correlation with global transverse stiffness E_{22} under mechanical loading. Figure 12 shows the comparison of transverse tensile strength S_{22}^+ values of the virtually cured RVEs and for those RVEs where cure-induced damaged has not been taken into account. It can be seen that for all the volume fractions and all σ_{cr} values considered, the RVEs that have cure-induced damage have noticeably lower strength values.

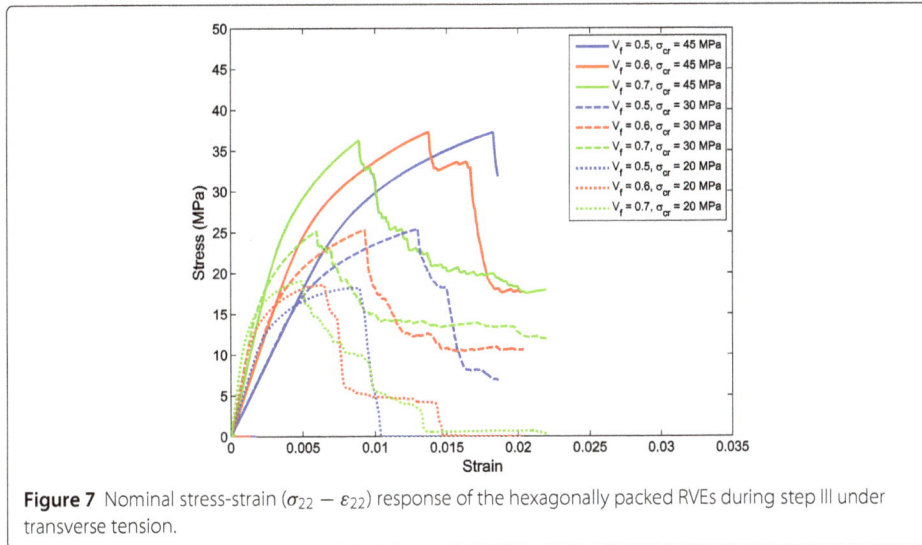

Figure 7 Nominal stress-strain ($\sigma_{22} - \varepsilon_{22}$) response of the hexagonally packed RVEs during step III under transverse tension.

Randomly packed fiber RVEs

Although the RVEs discussed thus far are idealized hexagonally packed geometries, they do not represent a RVE of a realistic FRPC sample. In realistic FRPCs, the fibers are randomly distributed which give rise to several matrix-rich pockets. It would be instructive to understand the severity of the cure-induced damage on the mechanical response, as a function of the randomness in fiber position in an RVE. Eight renditions of square FRPC RVEs with randomly distributed fibers are analyzed in this section. The distribution of fibers within the RVEs was done manually, in that the fibers were arbitrarily placed within the square RVE boundary. The fiber volume fraction (V_f) in all these renditions is chosen to be 0.55. These RVEs are shown in Figure 13. Few strategies to generate random RVEs may be found in the studies of Melro et al. [23], Yang et al. [24], and Vaughan and McCarthy [25]. Recently, using a heuristic random microstructure algorithm, Romanov et al. [26] have generated RVEs that are statistically well correlated with real FRPC RVEs.

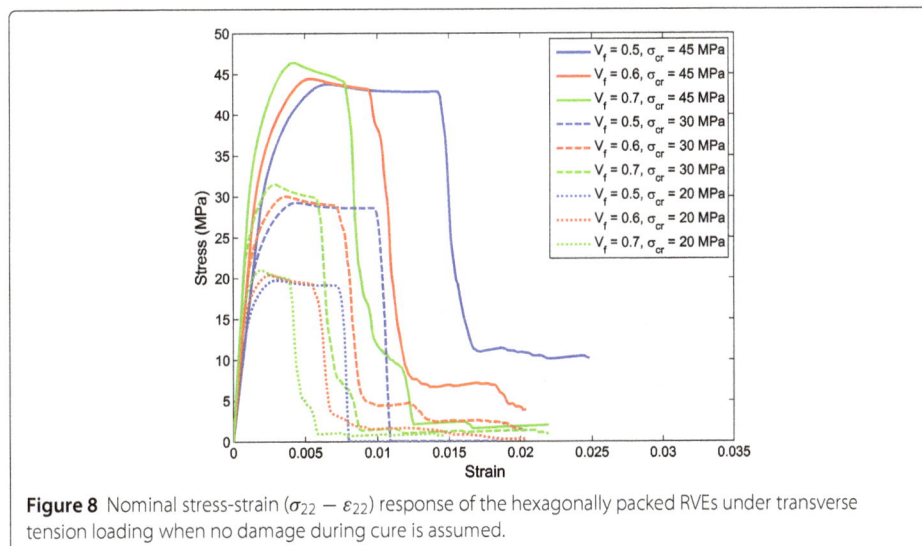

Figure 8 Nominal stress-strain ($\sigma_{22} - \varepsilon_{22}$) response of the hexagonally packed RVEs under transverse tension loading when no damage during cure is assumed.

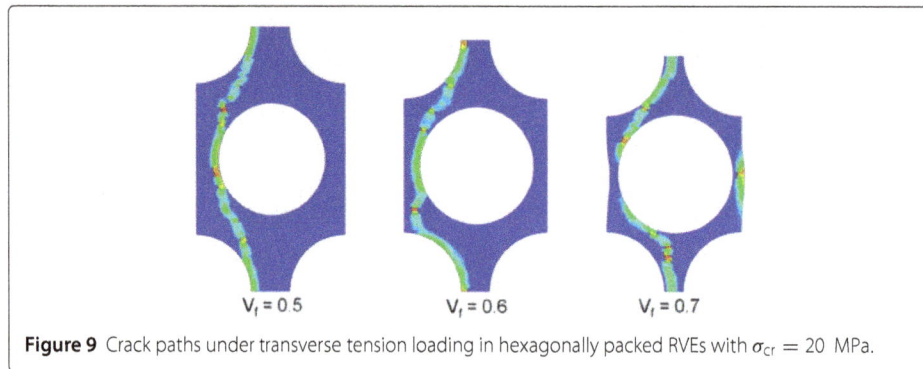

$V_f = 0.5$ $V_f = 0.6$ $V_f = 0.7$

Figure 9 Crack paths under transverse tension loading in hexagonally packed RVEs with $\sigma_{cr} = 20$ MPa.

The cure cycle, fiber, and matrix properties are similar to those used in the aforementioned study with hexagonally packed RVEs. A preliminary analysis on the mesh size has been conducted on a random RVE to establish that important features such as the stiffness reduction factor D at the end of step II and crack path at the end of step III are both mesh insensitive for the range of element sizes analyzed in this study. Figure 14 shows three different levels of refinement for a random packed case study. Results in terms of the factor D and crack path are shown in the top and bottom images of Figure 15, respectively. It can be seen that the spatial distribution of D and the two-piece failure paths are fairly consistent between the three meshes considered, thus establishing mesh objectivity.

Next, the eight random fiber RVEs shown in Figure 13 are cured with the crack band model (with critical fracture stress $\sigma_{cr} = 30$ MPa during cure) prescribed to capture any local matrix damage during cure. Periodic boundary conditions are enforced. The stiffness reduction factor D for these renditions is shown in Figure 16. Each of the RVEs exhibits nonhomogeneity in the contour for D. Note that on account of inherent randomness in fiber packing, there is no symmetry in the RVE at the start of cure. Matrix region areas that are surrounded by closely packed fibers are seen to exhibit higher stresses. This introduces stress gradients in different parts of the RVE during cure. Then, damage initiates at locations where the tensile stresses attain the critical fracture strength σ_{cr}. Therefore, different regions in the virtually cured matrix end up having nonhomogeneous stiffness values owing to different regions damaging differently during cure. These RVEs are next subjected to transverse tension loading along the 2-direction. The nominal stress-strain response is shown in Figure 17. It can be seen that the response is fairly linear during initial stages of loading (for nominal strain $0 \leq \varepsilon \leq 0.0025$). Beyond $\varepsilon > 0.0025$,

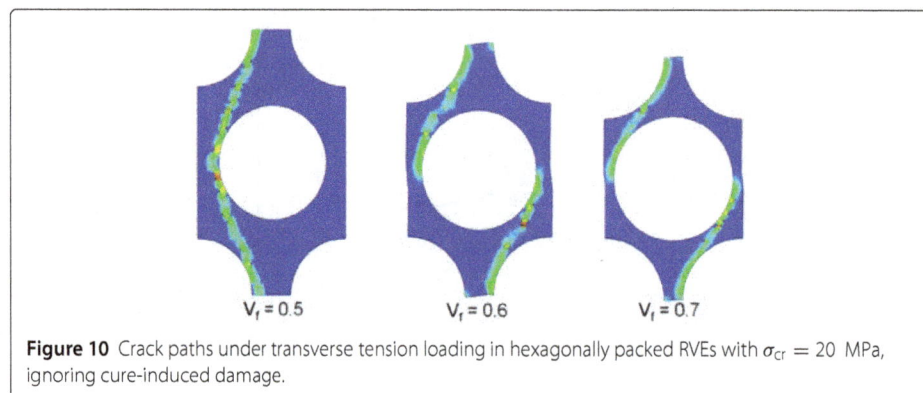

$V_f = 0.5$ $V_f = 0.6$ $V_f = 0.7$

Figure 10 Crack paths under transverse tension loading in hexagonally packed RVEs with $\sigma_{cr} = 20$ MPa, ignoring cure-induced damage.

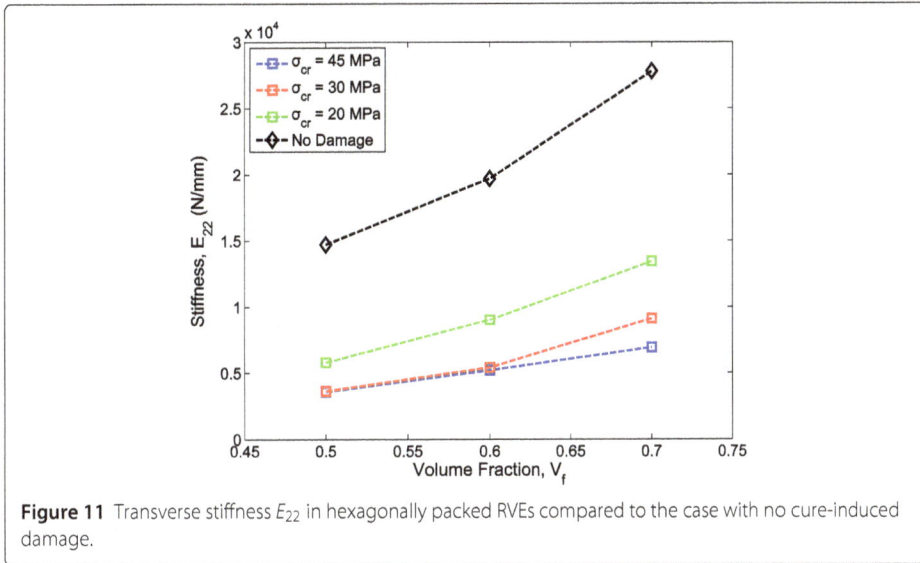

Figure 11 Transverse stiffness E_{22} in hexagonally packed RVEs compared to the case with no cure-induced damage.

there is nonlinear response followed by a peak value between strains of about $\varepsilon \approx 0.008$. Past the peak, the response is like that of a brittle solid, i.e., there is a drastic drop in stress due to two-piece matrix tensile failure. The two-piece crack paths for each of the random RVEs are shown in Figure 18. In some of the RVEs, the crack path is more tortuous than others. It is interesting to look at RVE #5, where there is a prominent and continuous *matrix-rich region* transverse to the loading direction. The two-piece crack path in this RVE at the end of step III is seen to propagate along a zone that has fibers that are more closely packed, which is away from the matrix-rich region. Similar observation holds for RVEs #1, #7, and #9 which have prominent but isolated matrix-rich regions. The matrix which is in a region where fibers are closely packed encounters higher stresses during curing as well as during mechanical loading and is more susceptible to cracking. Thus, cracks tend to initiate and propagate from such sites. The global stress-strain response of the random RVEs when cure-induced damage is not considered is shown in Figure 19.

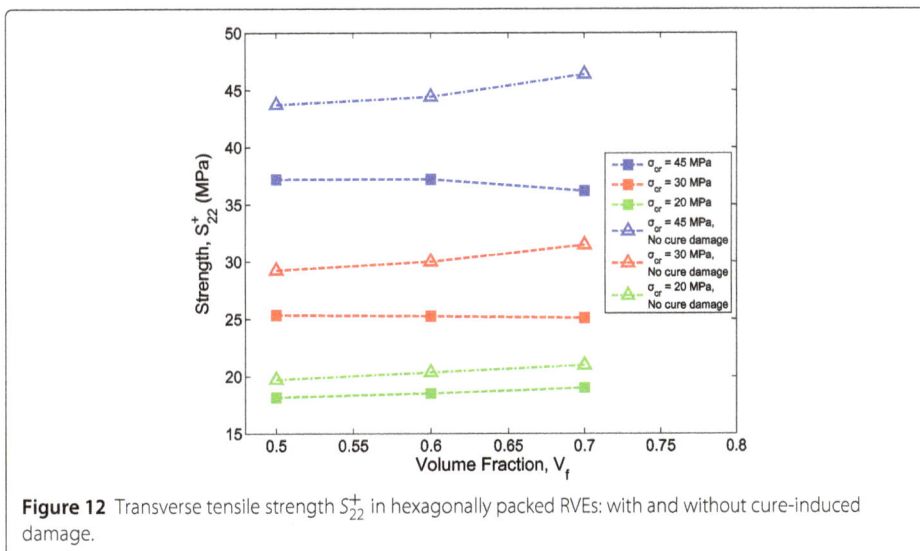

Figure 12 Transverse tensile strength S_{22}^{+} in hexagonally packed RVEs: with and without cure-induced damage.

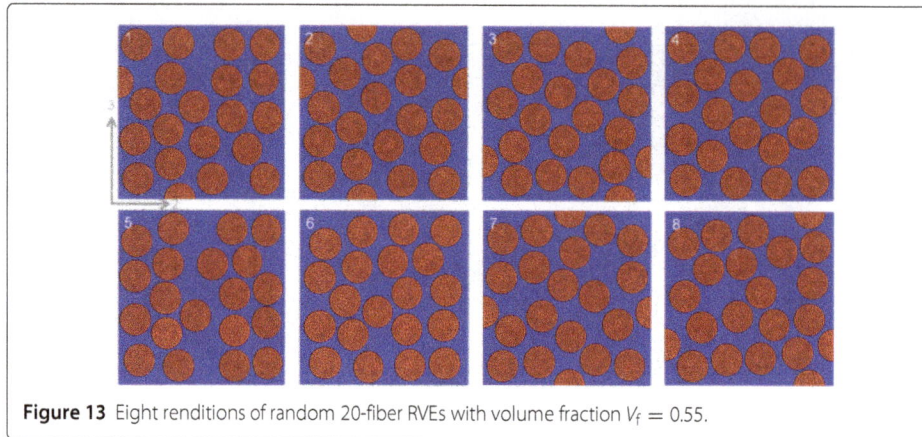

Figure 13 Eight renditions of random 20-fiber RVEs with volume fraction $V_f = 0.55$.

Here, the initial region is fairly linear (for nominal strain $0 \leq \varepsilon \leq 0.002$). However, the stiffness values in the initial region are identical compared to the RVEs with cure-induced damage, where there is a larger spread of the initial stiffness value owing to nonuniform stiffness distribution in the damaged matrix. There is some nonlinearity in the response which is much lesser than that seen in the RVEs when cure-induced damage was considered. A peak is attained beyond which the stress value plateaus momentarily followed by a drastic drop in stress. Finally, the strength values of the two cases (*with* and *without* cure-induced damage) are shown in Figure 20. For each of the RVEs with no cure-induced damage, the transverse tensile strengths are higher, i.e., mean $S_{22}^+ = 31.2$ MPa compared to mean $S_{22}^+ = 25.75$ MPa for RVEs with cure-induced damage. Moreover, the scatter in strength values in the RVEs with cure-induced damage is much larger than when no cure-induced damage is taken into account.

In the foregoing sections, we have established that damage in the matrix during cure results in a lower transverse strength value in the virtually cured hexagonally packed RVEs as well as virtually cured randomly packed RVEs. Thus, a comparison of the transverse tensile responses of these two types of RVEs is in order. In hexagonally packed RVEs having a fixed value of critical tensile fracture stress σ_{cr}, the transverse strength S_{22}^+ did not seem to vary with volume fractions ranging between 0.5 and 0.7 (see Figure 12). However, in randomly packed RVEs, the S_{22}^+ values exhibited significant scatter (see Figure 20). When compared to the case with no damage during cure, the mean transverse tensile strength reduction for hexagonally packed RVEs (15% reduction) and that for randomly packed RVEs (17% reduction) are similar. The worst strength reduction in randomly packed RVE is 25% for RVE #8. This shows that fiber packing within an RVE has an effect on the strength of the cured RVE with volume fraction held fixed. The effect of

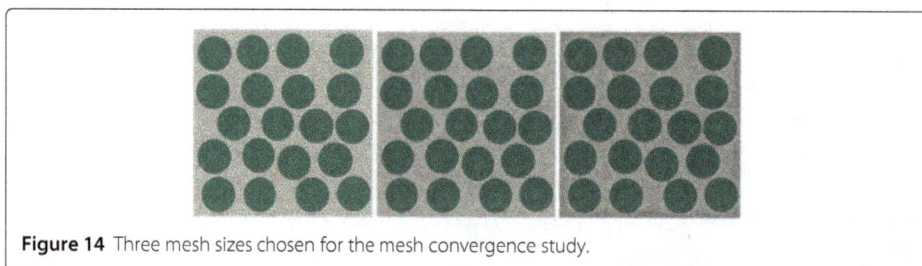

Figure 14 Three mesh sizes chosen for the mesh convergence study.

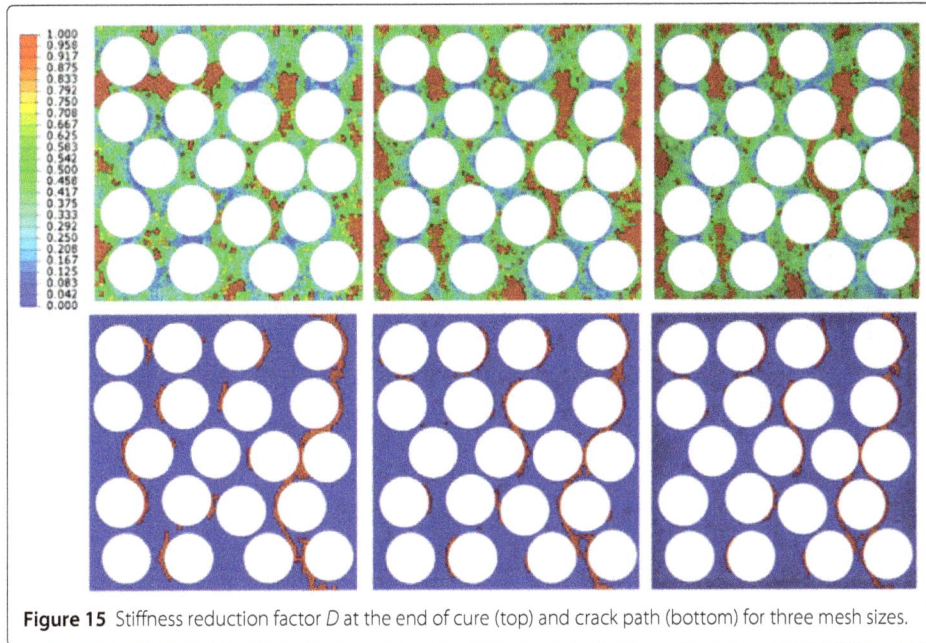

Figure 15 Stiffness reduction factor D at the end of cure (top) and crack path (bottom) for three mesh sizes.

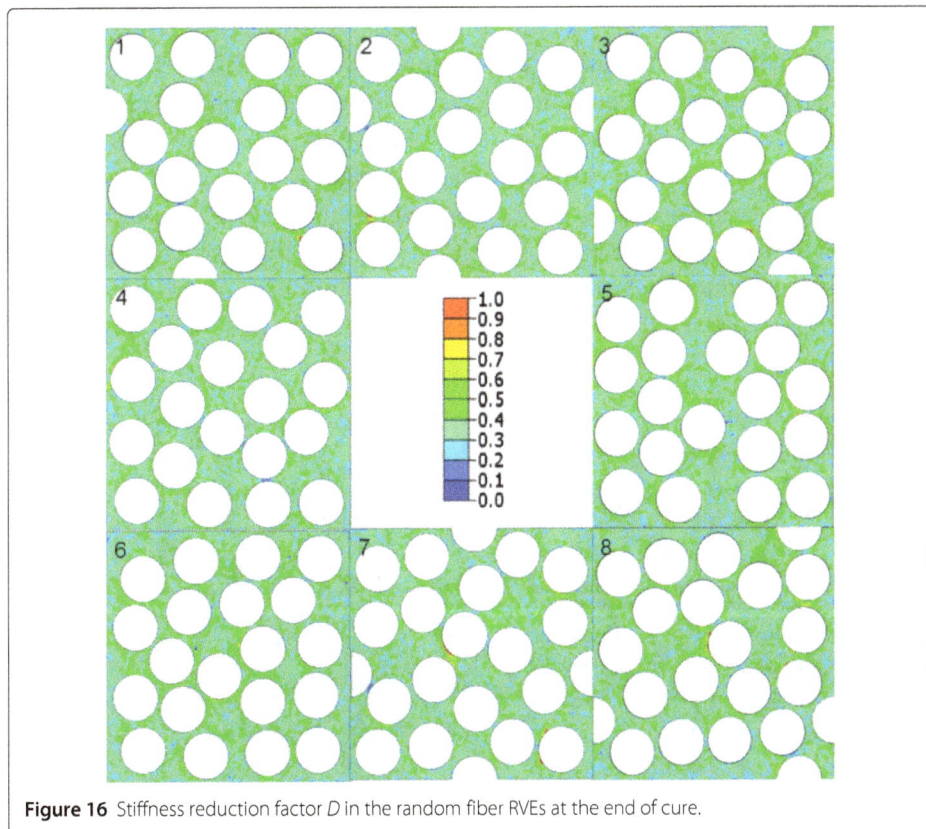

Figure 16 Stiffness reduction factor D in the random fiber RVEs at the end of cure.

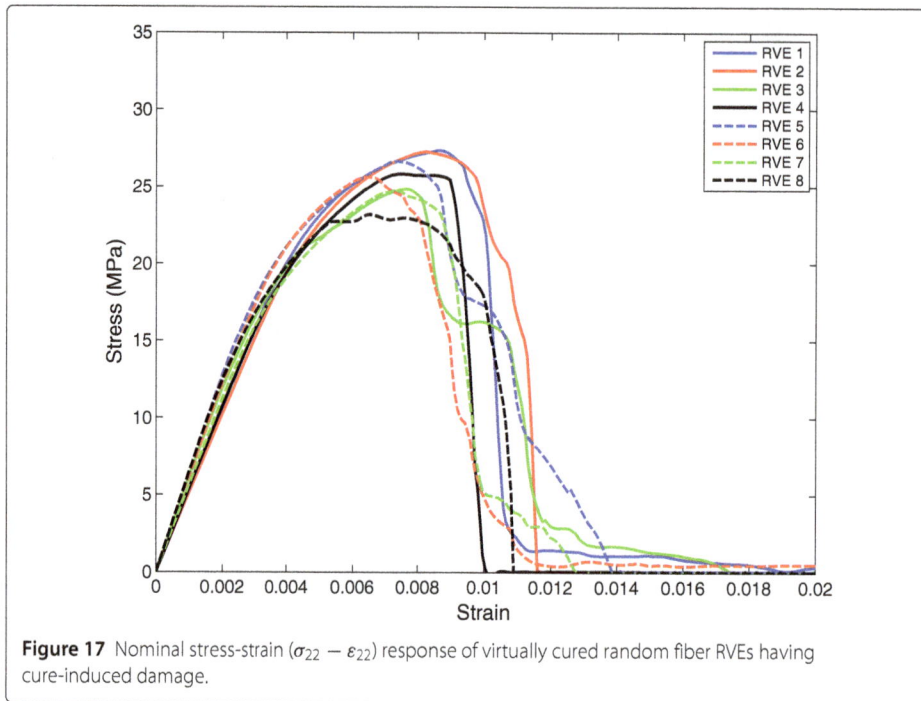

Figure 17 Nominal stress-strain ($\sigma_{22} - \varepsilon_{22}$) response of virtually cured random fiber RVEs having cure-induced damage.

microstructural randomness on mechanical response is detailed in the monograph by Ostoja-Starzewski [27].

Conclusions

The influence of cure on the mechanical response of virtually cured fiber-reinforced polymer matrix RVEs has been studied using a previously reported network model proposed by Heinrich et al. [15] in conjunction with the Bažant-Oh crack band model. Transverse tensile strength (S_{22}^+) and transverse stiffness (E_{22}) of these virtually cured RVEs were compared with those when no cure-induced damage was taken into account. These two quantities were calculated from the nominal stress-strain response of the virtually cured RVEs subjected to tensile loading along the transverse direction. Damage during cure is seen to reduce both stiffness and strength of the cured RVEs. Moreover, it is seen that even though fiber volume fraction is held fixed, the transverse strength of the virtually

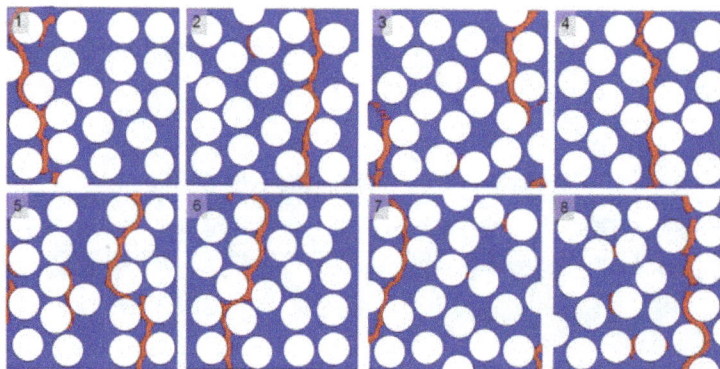

Figure 18 Two-piece crack paths in the randomly packed RVEs under transverse tension loading.

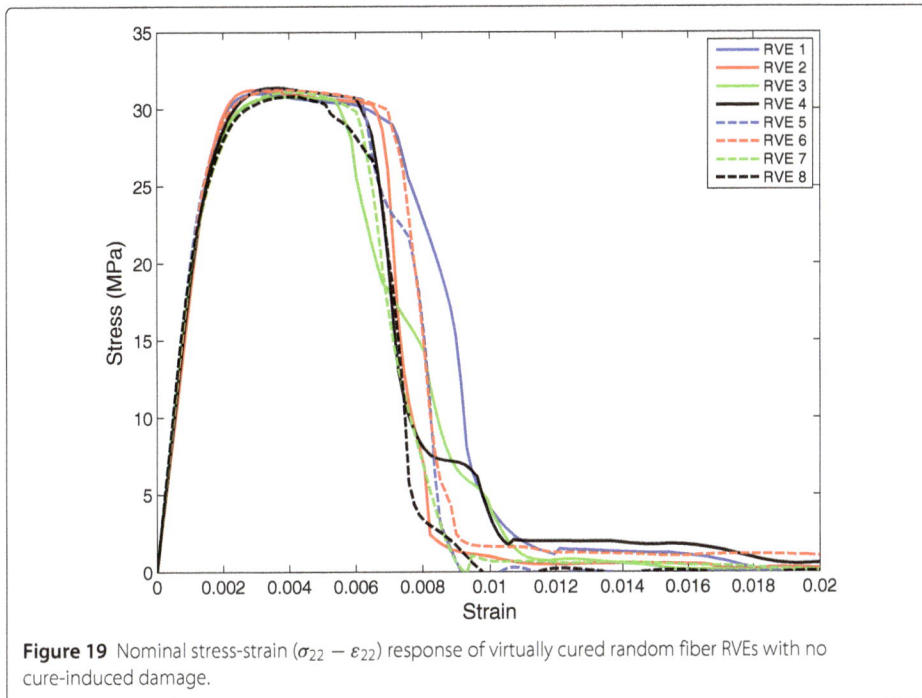

Figure 19 Nominal stress-strain ($\sigma_{22} - \varepsilon_{22}$) response of virtually cured random fiber RVEs with no cure-induced damage.

cured RVEs depend on the fiber packing. Also, the scatter in transverse strength values for RVEs with cure-induced damage is appreciably higher than in the case where cure-induced damage has been neglected. Since fiber packing is seen to influence strength of the cured RVE of constant fiber volume fraction, the study of correlating strength values with some metric associated with fiber packing is delegated to a future study. Using the approach described in this paper, several cure cycles can be considered and eventually tailored to arrive at an optimal cure cycle to reduce damage in the microstructure during the curing process, leading to superior mechanical strength and stiffness of the cured product.

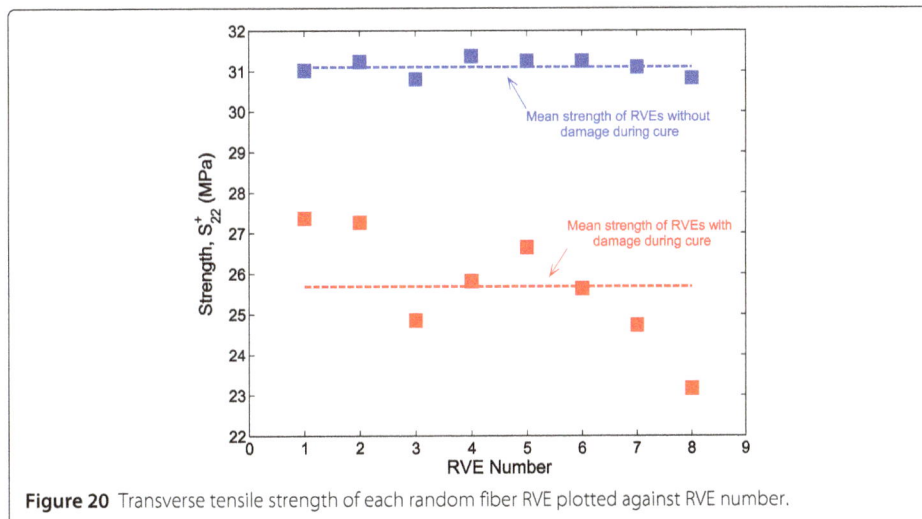

Figure 20 Transverse tensile strength of each random fiber RVE plotted against RVE number.

Competing interests

The authors declare that they have no competing interests.

Authors' contributions

RD and MM conducted the computational analysis and drafted the manuscript. AW conceived the study, its technical foundation, supervised its design and coordination, and contributed to writing the final version. All authors read and approved the final manuscript.

Authors' information

The authors are affiliated with William E. Boeing Department of Aeronautics and Astronautics, University of Washington, Seattle since 1 January 2015.

Acknowledgements

The authors thank Dr. Pascal Meyer and Dr. Christian Heinrich, of the Aerospace Engineering Department at the University of Michigan, Ann Arbor, and Prof. Pavana Prabhakar, Mechanical Engineering Department, University of Texas, El-Paso, for support with the user-defined subroutines used in the present work. The support of the Department of Aerospace Engineering, University of Michigan, Ann Arbor and the William E. Boeing Department of Aeronautics and Astronautics at the University of Washington, Seattle, is gratefully acknowledged.

References

1. Plepys AR, Farris RJ (1990) Evolution of residual stresses in three-dimensionally constrained epoxy resins. Polymer 31(10):1932–1936
2. Plepys AR, Vratsanos MS, Farris RJ (1994) Determination of residual stresses using incremental linear elasticity. Composite Struct 27(1-2):51–56
3. Merzlyakov M, McKenna GB, Simon SL (2006) Cure-induced and thermal stresses in a constrained epoxy resin. Composites: Part A 37:585–591
4. Chekanov YA, Korotkov VN, Rozenberg BA, Dhzavadyan EA, Bogdanova LM (1995) Cure shrinkage defects in epoxy resins. Polymer 36:2013–2017
5. Rabearison N, Jochum C h, Grandidier JC (2009) A FEM coupling model for properties prediction during the curing of any epoxy matrix. Comput Mater Sci 45(3):715–724
6. Ahn J, Waas AM (2002) Prediction of compressive failure in laminated composites at room and elevated temperature. AIAA Journal 40(2):346–358
7. Song S, Waas AM, Shahwan KW, Xiao X, Faruque O (2007) Braided textile composites under compressive loads: modeling the response, strength and degradation. Composite Sci Technol No. 67:3059–3070
8. Kim K, Hahn H (1989) Residual stress development during processing of graphite/epoxy composites. Composites Sci Technol 36:121–132
9. Li M, Zhu Q, Geubelle PH, Tucker III CL (2001) Optimal curing for thermoset matrix composites: thermochemical considerations. Polymer Composites 22:118–131
10. Gopal AK, Adali S, Verijenko VE (2000) Optimal temperature profiles for minimum residual stress in the cure process of polymer composites. Composite Struct 48:99–106
11. White S, Hahn H (1993) Cure cycle optimization for the reduction of processing-induced residual stresses in composite materials. J Composite Mater 27:1352–1378
12. Halpin JC, Kardos JL (1976) Halpin-Tsai equations: a review. Polymer Eng Sci 16(5):344–352
13. Mei Y (2000) Stress evolution in a conductive adhesive during curing and cooling. Ph.D Thesis, University of Michigan
14. Mei Y, Yee AS, Wineman AS, Xiao C (1998) Stress evolution during thermoset cure. Mater Res Soc Symp Proc 515:195–202
15. Heinrich C, Alridge M, Wineman AS, Kieffer J, Waas AM, Shahwan KW (2012) Generation of heat and stress during the cure of polymers used in fiber composites. Int J Eng Sci 53:85–111
16. Kamal MR (1974) Thermoset characterization for moldability analysis. Polymer Eng Sci 14(3):231–239
17. Li C, Potter K, Wisnom MR, Stinger G (2004) In-situ measurement of chemical shrinkage of MY750 epoxy resin by a novel gravimetric method. Composites Sci Technol 64(1):55–64
18. Simulia (2012) Abaqus user manual, version 6.12. Dassault Systèmes, Providence, RI, USA
19. Bažant ZP, Oh B (1983) Crack band theory for fracture of concrete. Mater Struct 16(3):155–177
20. Jirasek M, Bažant ZP (2002) Inelastic analysis of structures. John Wiley & Sons, London and New York
21. Gonzalez C, Llorca J (2007) Mechanical behavior of unidirectional fiber-reinforced polymers under transverse compression: microscopic mechanisms and modeling. Composites Sci Technol 7:2795–2806
22. Xia Z, Zhang Y, Ellyin F (2003) A unified periodical boundary conditions for representative volume elements of composites and applications. Int J Solids Struct 40:1907–1921
23. Melro AR, Camanho PP, Pinho ST (2008) Generation of random distributions of fibers in long-fibre reinforced composites. Composites Sci Technol 68(9):2092–2102
24. Yang L, Ying Y, Ran Z, Liu Y (2013) A new method for generating random fiber distributions for fiber reinforced composites. Composites Sci Technol 76:14–20
25. Vaughan TJ, McCarthy CT (2010) A combined experimental-numerical approach for generating statistically equivalent fiber distributions for high strength laminated composite materials. Composite Sci Technol 70(2):291–297
26. Romanov V, Lomov SV, Swolfs Y, Orlova S, Gorbatikh L, Verpoest I (2013) Statistical analysis of real and simulated fibre arrangements in unidirectional composites. Composite Sci Technol 87:126–134
27. Ostoja-Starzewski M (2007) Microstructural randomness and scaling in mechanics of materials. Chapman and Hall-CRC 2007, Florida, USA

8

Credibly reaching a reliability target using a model initially constructed by expert elicitation

Lawrence E Pado

Correspondence:
lawrence.e.pado@boeing.com
The Boeing Company, 8900 Frost
Ave., Bldg 245 MC S245-1260, 63134
Berkeley, MO, USA

Abstract

The Defense Advanced Research Projects Agency Defense Science Office (DARPA/DSO) is sponsoring Open Manufacturing (OM), an initiative to develop new technologies, new computational tools, and rapid qualification to accelerate the manufacturing innovation timeline. Certification Methodology to Transition Innovation (CMTI), an OM program, has developed a methodology to quantify the effect of manufacturing variability on product performance to address the risk to cost and performance associated with failure to take manufacturing capability and material and fabrication/assembly variation into account early in the design process. An important aspect of this program is the use of Bayesian networks (BN) to evaluate risk. The BN is used as a graphical representation of the contributing factors that lead to manufacturing defects. The reliability of the final product is then analyzed using the contributing factors. There are many types of programs where there is little relevant data to support the probabilities needed to populate the BN model. This is very likely the case for new programs or at the end of long programs when obsolescence challenges servicing a product when original vendors are no longer in business. In these cases, probabilities must be obtained from expert opinion using a technique called expert elicitation. Even under objective 'Good Faith' opinions, the expert himself has a lot of uncertainty in that opinion. This paper details an approach to obtaining credible model output based on the idea of having a hypothetical expert whose unconscious bias influences the model output and discovering and using countermeasures to find and prevent these biases. Countermeasures include replacing point probabilities with beta distributions to incorporate uncertainty, 95% confidence levels, and using a multitude of different types of sensitivity analyses to draw attention to potential trouble spots. Finally, this paper uses a new technique named 'confidence level shifting' to optimally reduce epistemic uncertainty in the model. Taken together, the set of tools described in this paper will allow an engineer to cost effectively determine which areas of the manufacturing process are most responsible for performance variance and to determine the most effective approach to reducing that variance in order to reach a target reliability.

Keywords: Credibility; Expert elicitation; Confidence level shifting; Monte Carlo; Uncertainty quantification; Targeted testing; Unitized testing; Uncertainty reduction; Epistemic uncertainty; Reliability targets

Background

The Defense Advanced Research Projects Agency Defense Science Office (DARPA/DSO) is sponsoring Open Manufacturing (OM), an initiative to develop new technologies, new computational tools, and rapid qualification to accelerate the manufacturing innovation timeline. Certification Methodology to Transition Innovation (CMTI), one of the programs in the OM portfolio, has developed a methodology to quantify the effect of manufacturing variability on product performance to address the risk to cost and performance associated with failure to take manufacturing capability and material and fabrication/assembly variation into account early in the design process.

Motivation

The goal motivating this research is to first credibly ascertain the reliability of a product by including the effects of variability and defects in manufacturing as well as uncertainty in the environment. Note that credibility is a key requirement and is made more difficult when using expert elicitation to determine the value of model parameters used to calculate the reliability. Secondly, upon evaluation of the manufacturing process, it is very likely that product reliability will fall short of the desired target. The set of tools described in this paper will allow the engineer to cost effectively determine which areas of the manufacturing process are most responsible for performance variance and to determine the most effective approach to reducing that variance in order to reach a target reliability. These benefits will be explained in detail in the sections entitled 'Techniques to meet a target POF with a 95% confidence level' and 'Putting it all together - an example using credibility tools'.

An exemplar problem

An exemplar problem used to drive development of the framework was the manufacture of an out-of-autoclave composite panel stiffened with three hats (Figure 1 [1]). The hat-stiffened panel represents a design/manufacturing feature-based element or subcomponent in the traditional building block approach. The manufacturing steps called out in the fabrication work order serve as the initial basis for creating a Bayesian network (BN) [2] used to tailor risk with a quality control plan and to determine the probability of defects. The BN is used as a graphical representation of the contributing factors that lead to manufacturing defects. The reliability of the final product is then analyzed using the contributing factors.

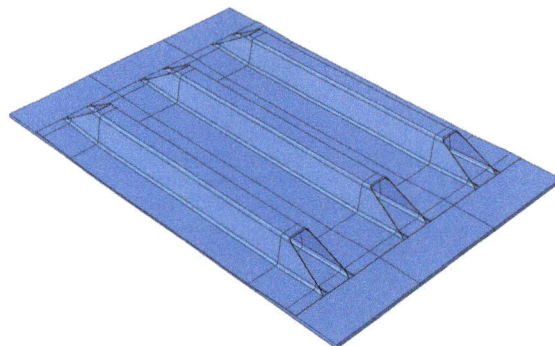

Figure 1 Isometric view of hat-stiffened panel (69 cm wide × 91 cm long).

The environmental condition that this part is to be evaluated against for the purposes of this paper is out-of-plane pull-off of a hat structure (Figure 2). It should be noted that this approach used multiple load cases as shown in Figure 2, but for simplicity, only the pull-off case will be discussed. As described in detail in reference [3], given a probabilistic environmental load, geometrical and material property variation, as well as probabilistic manufacturing defects, a probability of failure (POF) of the hat-stiffened panel can be calculated.

POF is partially a function of the probability of defects occurring during the manufacture of a part. A subset of all possible defects that can be introduced by the manufacturing process, tooling, etc., and that are thought to contribute to failure under acknowledged conditions, were identified. The focus of this work was on quantifying and reducing manufacturing defects. Although material and environmental variability were accounted for, examination of the costs and benefits of reducing the variability of those factors were beyond the scope of this research. They are however important factors and will be considered in future work. For the hat-stiffened panel, the defects recognized were wrinkles (nugget/noodle fiber waviness), noodle void/porosity/geometry, lower radius thickening, upper radius thinning, and top crowning. If there is any doubt as to if a defect can affect performance, it should be included in the analysis.

Once the defects of interest were identified, the step-by-step manufacturing process was analyzed to determine which steps or combinations of steps could possibly produce one or more of those defects. Additionally, options that could affect the probability of introducing defects were identified such as tooling choices, manufacturing capability levels, and manufacturing process alternatives.

At this point in the knowledge collection process, enough information was obtained to create the structure of a Bayesian network. If there is any doubt as to whether or not a factor can influence the probability of a defect, it should be included in the model and assigned a high uncertainty. Uncertainty assignment will be discussed later in this paper. A fragment of the complete BN is shown in Figure 3. The overall purpose of this Bayesian network is to calculate the probability of defects and the resulting probability of failure of the structure. For those readers familiar with process flows, it should be noted that the BN will not necessarily mimic this flow, but will instead be built to capture direct relationships between the process variables. The resultant probability of each defect (and thus the POF) is a function of any and all combinations of the choices that can go into the manufacturing process. The BN is used to determine POF with acceptable cost, or it can be used to find the cost optimal manufacturing process choices given a desired POF as described in [3].

In order for the network to calculate these total defect and POF probabilities, however, the probability that each individual manufacturing step can induce a defect

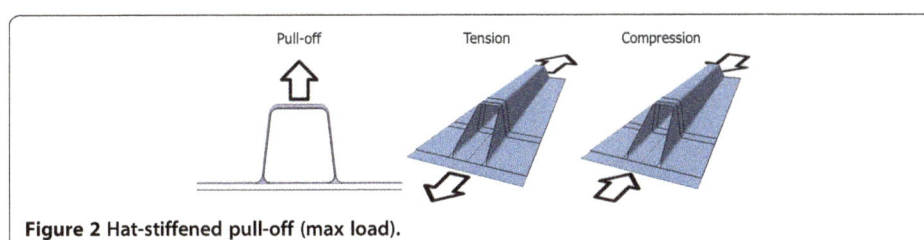

Figure 2 Hat-stiffened pull-off (max load).

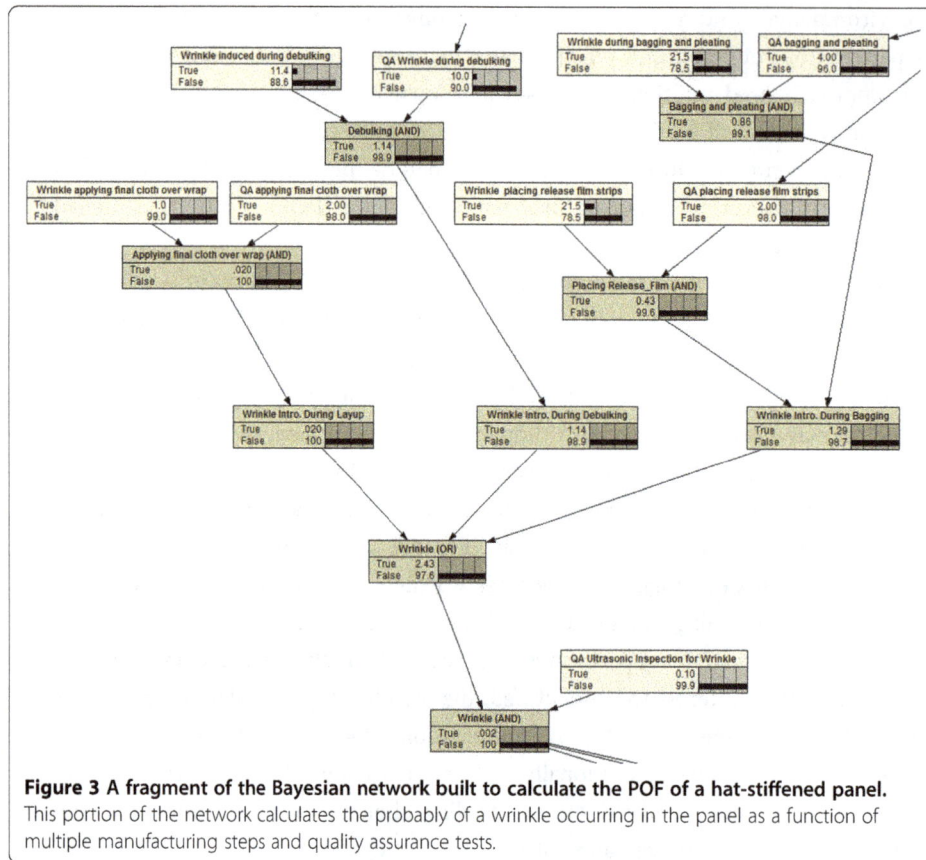

Figure 3 A fragment of the Bayesian network built to calculate the POF of a hat-stiffened panel. This portion of the network calculates the probably of a wrinkle occurring in the panel as a function of multiple manufacturing steps and quality assurance tests.

and that each individual quality assurance test can find it, if it exists, must be provided.

Methods

An approach to determining the credibility of models

For those programs where there is little relevant data to support the probabilities needed to populate the BN model, expert elicitation must be used to provide them. Even under objective 'Good Faith' opinions, the expert himself has a lot of uncertainty in that opinion. On a given day, the expert may even choose different probabilities than the one he had chosen earlier. Given that the output of the model has real-world consequences, possibly in terms of customer acceptance of the product, there may be bias when choosing probabilities - especially if the model has not shown that it can meet a target POF.

The chosen way to approach this issue is to make the potential bias explicit by figuring out the best way to modify model inputs to get a desired result. Once this methodology is known, the idea is to reverse engineer it to find countermeasures and establish the credibility of the model.

The expert's first approach - using and adjusting point probabilities

The expert's initial approach is to make estimates of the probabilities as objectively as possible. If the target POF is reached, then the expert is done. If not, he will find the point probabilities that the output is most sensitive to and adjust them as little as possible such that the target is reached. Note that point probabilities are probabilities that are assumed to be known with absolute certainty and are represented by a single scalar value. The reasoning behind this strategy is that adjusting the probabilities that do not have much of an effect would require large unrealistic changes to have a significant effect.

Derivative-based approach to sensitivity

The typical way to perform a sensitivity analysis [4] is to calculate the partial derivative of the output with respect to a model parameter as shown in Equation 1.

$$S_{D_i} = \frac{\delta Y}{\delta X_i} \tag{1}$$

where S_{D_i} is the derivative-based sensitivity measure with respect to parameter i, Y is the model output of interest, and Xi is the model parameter i.

In practice for a Bayesian network model, this sensitivity analysis is performed numerically. The output of interest for the exemplar problem is the POF of the three-hat-stiffened panel under a pull-off load. To calculate the sensitivity of the POF to each node's point probability of either inducing a defect or for QA nodes, the probability of missing a defect if it exists, Equation 2 is used:

$$S_{D_i} = \frac{\Delta \text{POF}}{\Delta P_i} \tag{2}$$

where S_{D_i} is the derivative-based sensitivity measure with respect to probability i, POF is the probability of failure of the three-hat panel under pull-off load, and P_i is the probability of node i inducing a defect or failing to detect a defect.

With this technique in hand, the expert would calculate the sensitivity of POF to changing the probabilities within every manufacturing step or QA test node of the network. After sorting from most sensitive to least sensitive, he could calculate exactly how much to change the top few nodes to reach his target POF.

Quantifying uncertainty in the model probability parameters - the beta distribution

Although one possible countermeasure to this would be to perform a similar sensitivity analysis and use the findings to focus scrutiny on the highest sensitivity nodes, a bigger issue is that the point probabilities express complete certainty in the model values. The first countermeasure is to note that it is not possible for an expert to have absolute certainty in the values used as model parameters. Even with hard data, there is some uncertainty as to the exact value. The conclusion to be drawn, then, is that each model parameter, each of which is a probability, should be represented by a probability density function, ideally one that has zero as a lower bound and one as a higher bound due to the fact that probabilities by definition are always between 0 and 1. The beta distribution is such a distribution and is useful for representing binary success/failure problems, specifically the proportion of successes or failures that would be expected over time. Mathematically, the beta distribution is represented by Equation 3 [5].

$$beta(a, b) = \frac{1}{\beta(a, b)} p^{a-1}(1-p)^{b-1} \tag{3}$$

where p = proportion of flaws $0 <= p <= 1$, a = number of flawed examples (positive number), b = number of flawless examples (positive number), and β = the beta function (not to be confused with the beta distribution).

The beta distribution is quite flexible and can represent uniform ($a = 1$, $b = 1$), ramp, symmetrical, and asymmetrical distributions. It is simple to implement Bayesian learning using newly introduced data using this distribution. For each step in the manufacturing process, if a flaw occurs during that step, the parameter 'a' merely needs to be incremented by one. Likewise, if no flaw is introduced during that step, the parameter 'b' should be incremented by 1. For quality assurance (QA) tests, if a QA test does not miss a flaw that exists, then 'b' should be incremented. If the test misses a flaw that exists, 'a' should be incremented.

Using expert opinion elicitation to determine the parameters of the beta distribution

The goal of expert opinion elicitation is to determine the parameters of beta distribution such that it accurately represents the expert's opinion about the most likely value of the probability to be specified (the mode) as well as his uncertainty in that opinion. In this methodology, this is accomplished by having the expert express his uncertainty in terms of the number of samples he has seen. The following basic example will build the readers intuition about this process.

Upon purchasing an unlabeled trick coin received at a magic shop, it is desirable to determine the characteristics of that coin. The bin the coin was stored in noted only that the coin could be weighted to always come up heads (a two-headed coin), always come up tails (a two-tailed coin), or anything in between. This information from the bin represents prior knowledge that the coin can be weighted to any degree. This can be represented by the uniform distribution denoted beta (1, 1). Beta (1, 1) is also known as the non-informative prior distribution. Figure 4A shows a beta distribution in which every weighting is equally likely. To gather more data, the coin is flipped three times and this results in three heads. Clearly, while this coin is definitely not weighted to always come up tails, it is certainly possible that this is a fair coin that has been weighted to come up heads or tails an equal number of times, but it seems more likely that this coin is weighted towards heads. Figure 4B plots that beta distribution after adding in this new data as beta (1, 1 + 3 heads).

Finally after 30 flips, and obtaining 30 heads in a row, it is clear that this is not a fair coin but is very highly weighted towards coming up heads. The updated beta distribution is shown in Figure 4C. Note that even after 30 flips, it is not a sure thing that a heads result will always be obtained. Also note that the distribution is getting narrower and narrower representing an increase in the certainty of the coin's weighting value.

Using the above example as an intuitive example of the meaning of 'samples seen', the expert can be asked to provide a level of uncertainty in terms of samples seen. For the exemplar problem, what proportion of wrinkles has been observed during the debulking process (i.e., the probability of a wrinkle occurring during the debulking process)? Note that this is represented by the node named 'Wrinkle induced during debulking' in Figure 3.

Given these two pieces of information, the most likely value (mode) and the uncertainty in terms of samples seen, the parameters of the beta distribution that meets these requirements are as follows [6]:

$$a = \text{mode} \times (k-2) + 1 \tag{4}$$

$$b = (1-\text{mode}) \times k-2 + 1 \tag{5}$$

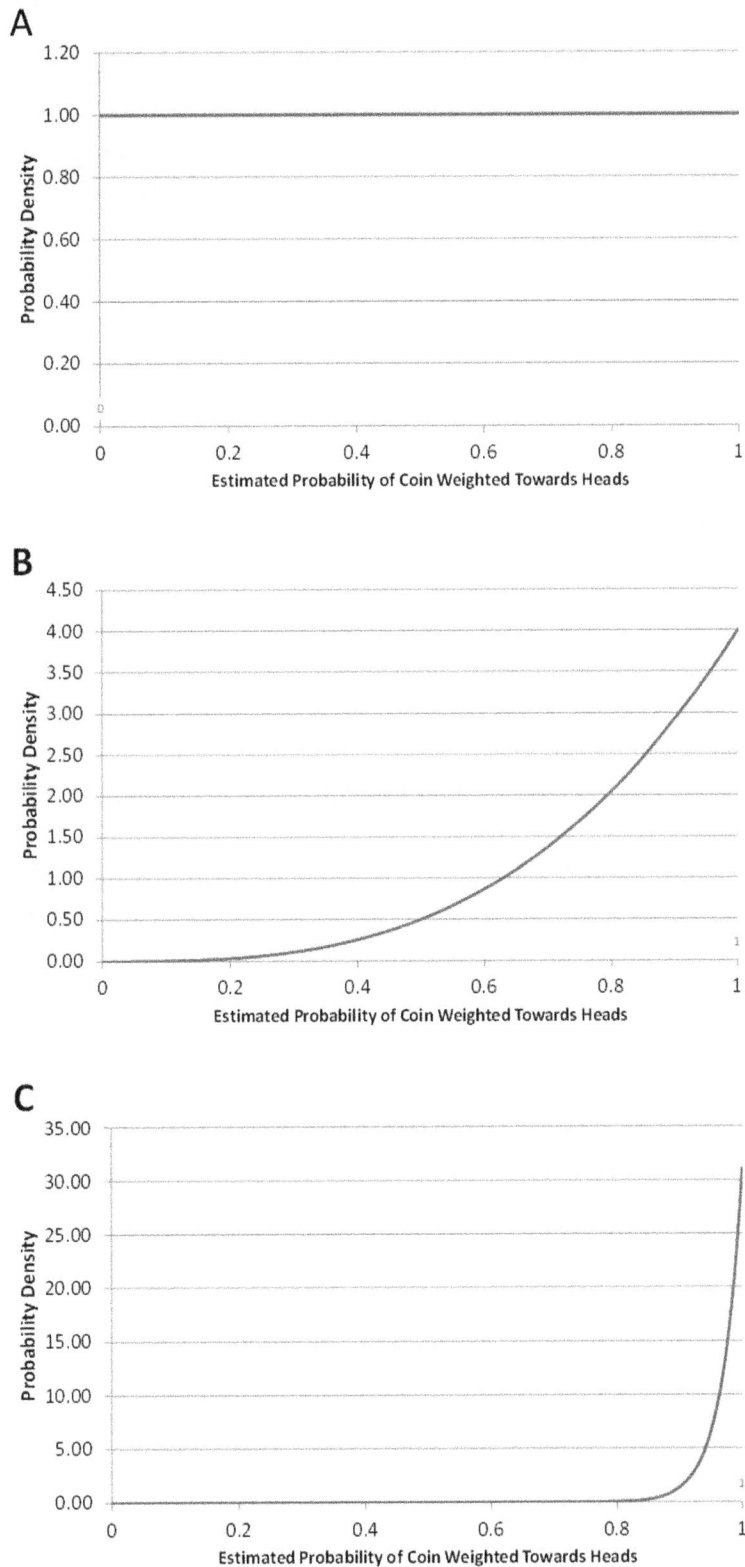

Figure 4 The non-informative beta distribution updated with additional data. (A) The beta distribution beta (1,1), which represents the non-informative state also known as the uniform. **(B)** The beta distribution beta (1,4), which represents uniform distribution updated by getting three heads in three flips. **(C)** The beta distribution beta (1,31), which represents uniform distribution updated by getting 30 heads in thirty flips.

where a = beta distribution parameter expressing the number of flawed examples, b = beta distribution parameter expressing the number of flawless examples, mode = the most likely probability of defect, and k = the expert's confidence in the estimate expressed in terms of equivalent prior sample size (minimum 2).

In the debulking example above, the expert provided a most likely value of 0.11 with a sample size of 120. Figure 5 shows how a point probability of 0.11 is now represented by a beta distribution with parameters of beta (13.98, 106.02). Note that in this example, the expert is implying that the range of likely values for that probability is approximately between 0.04 and 0.22.

This process is continued for every node in the network until every parameter of the network is represented by a beta distribution.

Determining a 95% confidence value on the model output using Monte Carlo methods

Now that all of the model parameters have been replaced by beta distributions, the distribution of the model output can be computed using Monte Carlo methods [7]. Specifically, a 95% confidence value [8] can be calculated on the model output, which for the exemplar case is POF. Figure 6 provides intuition about the 95% confidence level (CL). As shown in the figure, it is the value at which there is a 95% chance that the true value is less than or equal to the CL. As can be seen, this is a quite conservative estimate since the most likely value (peak or mode) of the shown distribution is much smaller.

A Monte Carlo analysis entails pulling a single sample from each distribution within the model, populating the model with these new samples, and then running the model to get a single answer. This process is then repeated thousands of times to collect enough data to establish the distribution of the output parameter of interest such as POF and enables calculating its CL.

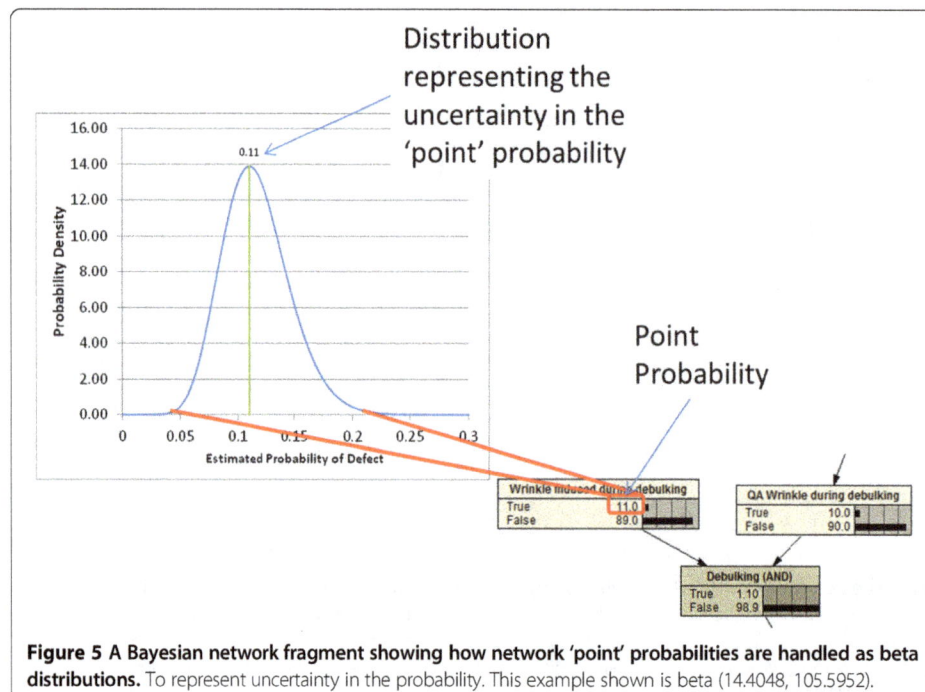

Figure 5 A Bayesian network fragment showing how network 'point' probabilities are handled as beta distributions. To represent uncertainty in the probability. This example shown is beta (14.4048, 105.5952).

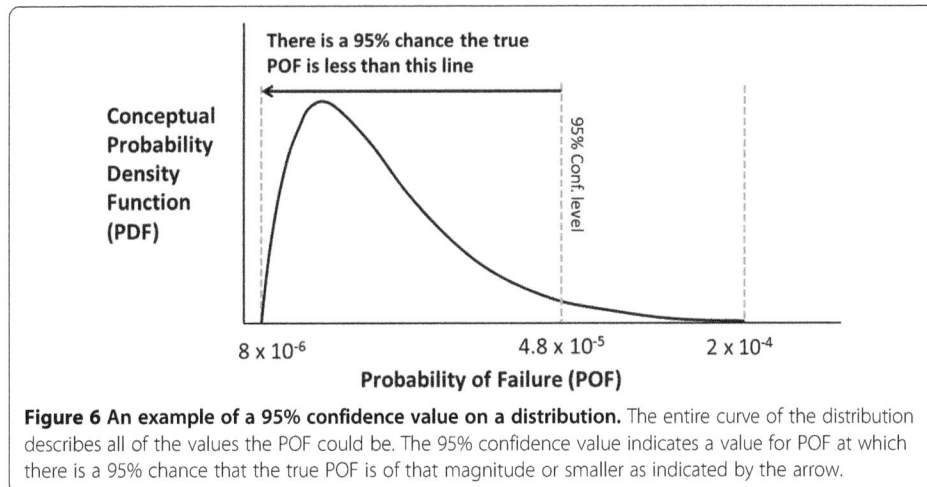

Figure 6 An example of a 95% confidence value on a distribution. The entire curve of the distribution describes all of the values the POF could be. The 95% confidence value indicates a value for POF at which there is a 95% chance that the true POF is of that magnitude or smaller as indicated by the arrow.

Finding the 95% CL using histogram data is a straightforward process involving sorting the POF data from lowest value to highest value and then selecting the value for which 95% of the data is that value or smaller.

The expert's second approach - adjusting mode and certainties of PDFs

Using probability distributions instead of point probabilities and instituting a 95% confidence level to introduce conservatism is a good first step for establishing credibility. The expert's bias, if it exists, may now show up as a lower mode value or as a higher level of certainty in that mode value. This type of bias must be detected if it is to be countered.

Countermeasure to the expert's second approach - sensitivity analysis on the mode with respect to POF

The countermeasure to the impact of the expert's mode and certainty selection is to perform a sensitivity analysis of the 95% confidence level to changes in the mode.

There are two conditions that have to be met before the 95% confidence level shows significant sensitivity to the mode of a node:

- The node has to have already been shown to be important through the use of sensitivity analysis. If the output has no sensitivity to the node, then the mode of the node is inconsequential.
- The sensitivity of 95% CL to a mode increases as the certainty in the value of probabilities increases. As discussed above, for beta distributions, certainty is a function of the number of samples expressed. A sample size of 2 will result in no sensitivity to mode with the sensitivity increasing as the number of samples increases.

Figure 7 illustrates these concepts. In this figure, a beta distribution's mode is doubled from 0.01 to 0.02 under 7A, a condition of high uncertainty (samples = 10) and 7B, a condition of low uncertainty (samples = 1,000). Under the condition of high uncertainty, a mode change has very little effect on the basic shape of the distribution. Under the condition of low uncertainty, the two distributions are much more distinct.

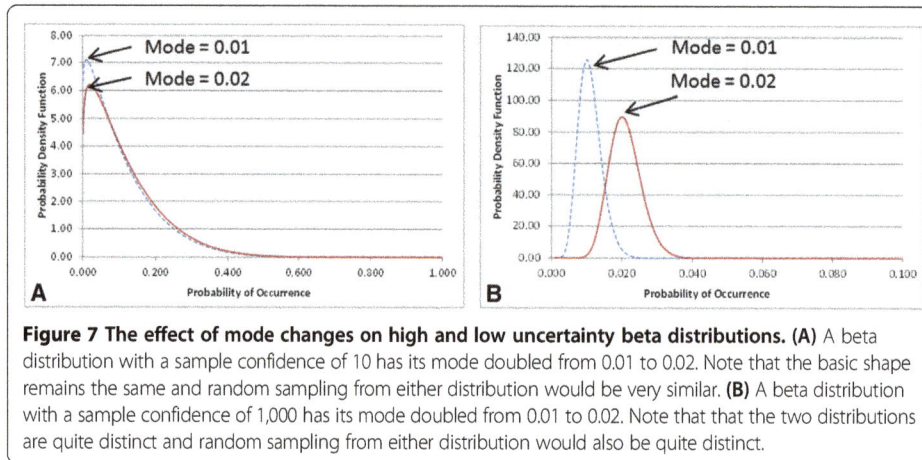

Figure 7 The effect of mode changes on high and low uncertainty beta distributions. (A) A beta distribution with a sample confidence of 10 has its mode doubled from 0.01 to 0.02. Note that the basic shape remains the same and random sampling from either distribution would be very similar. **(B)** A beta distribution with a sample confidence of 1,000 has its mode doubled from 0.01 to 0.02. Note that that the two distributions are quite distinct and random sampling from either distribution would also be quite distinct.

Mathematically, the sensitivity of 95% CL to Δmode is measured as:

$$S_{95\%CL_i} = \frac{\Delta 95\%CL}{\Delta Mode_i} \tag{6}$$

Where:

where $S_{95\%CL_i}$ = the sensitivity of the 95% confidence level of the model output of interest to a change in mode of node i, $\Delta 95\%$ CL = the change in the 95% confidence level due to a change in mode of node i, and $\Delta Mode_i$ = the change in mode of node i.

The following process can be used to calculate $S_{95\%CL_i}$

- Calculate the baseline 95% CL by running a Monte Carlo analysis on the baseline model.
- Choose a node i.
- Increase the mode of node i by a delta value.
- Calculate the new 95% confidence level by running a Monte Carlo analysis.
- Calculate the sensitivity as per Equation 6.
- Repeat this process for each node i of interest.

Table 1 shows the results of this process on three nodes for an illustrative exemplar.

As shown in Table 1, the credibility of the mode value in the first two nodes listed in the table is very high as even a very large change in the mode value would have very little effect on the 95% CL POF. In fact, the mode value of node 'QA test finds debulking wrinkle if it exists' has absolutely no effect on the value of 95% CL POF due to having the maximum uncertainty in its value. It should be noted, however, that the uncertainty in the mode value has a large impact on the variance of the output, as will be discussed in more detail below. The final node, 'UltraSonic inspection finds wrinkle,' with a sensitivity of 0.01453, indicates that it would have been possible for the expert to significantly change the 95% CL POF by changing the mode. More specifically, the 95% CL changes by 1.45×10^{-5} for every 0.001 change in the mode. This means, for example, that if the mode was originally 0.008 and the expert lowered it to 0.001, the 95% CL would have been $7 \times 1.45 \times 10^{-5}$ higher or 1.5×10^{-4} instead of the reported 4.8×10^{-5}

Table 1 Sensitivity calculations of 95% CL to a change in mode for an illustrative exemplar

Node i	NumSamples	Old_mode	New_mode	Delta_mode	95POF	New95POF	Delta_POF	Sen
Wrinkle introduced during debulking	120	0.114	0.227	0.114	4.80×10^{-5}	8.07×10^{-5}	3.26×10^{-5}	0.00029
QA test finds debulking wrinkle if it exists	2	0.100	0.200	0.100	4.80×10^{-5}	4.80×10^{-5}	0.00E+00	*0.00000*[a]
UltraSonic inspection finds wrinkle	1,000	0.001	0.002	0.001	4.80×10^{-5}	6.25×10^{-5}	1.45×10^{-5}	*0.01453*[b]

[a]Note that the sensitivity of 95% CL POF is zero for the mode of node 'QA test finds debulking wrinkle if it exists'. This is due to the mode having maximum uncertainty (samples = 2) as discussed previously; [b]the sensitivity of the mode of node 'UltraSonic inspection finds wrinkle' is fairly high at 0.01453. Thus, for every .001 change in mode, 95% CL POF changes by 1.45×10^{-5}. This effectively means that if the expert started with a mode of .008 and reduced it down to 0.001, a magnitude change in 95% CL POF would have occurred.

from Table 1. The consequence of this observation is that the expert should be required to provide documented proof of the 1,000 sample size or else he should be required to reduce his reported sample size.

Techniques to meet a target POF with a 95% confidence level

With these procedures in place, the expert may find that it is not possible to meet the target value of product reliability (i.e., a low enough probability of failure). What guidance can be provided to the expert to cost effectively increase reliability?

The goal is to raise the 'certain' reliability cost-effectively. The word 'certain' here is used to indicate that a reported low reliability may be due in part to a lack of process knowledge, while the other portion is due to variability in the manufacturing process coupled with a lack of suitable quality assurance tests. These ideas are captured by the following two types of uncertainty [9]:

- *Aleatory* variability is the natural randomness in a process. Aleatory uncertainty cannot be reduced thru data collection. For example, the knowledge of what number will turn up on a six-sided die. This type of uncertainty can be reduced through better process control and through quality assurance testing. In the die analogy, this is equivalent to reducing the number of sides on the die or weighting the die to come up favorably.
- *Epistemic* uncertainty is the scientific uncertainty in the model of the process. It is due to limited data and knowledge. This uncertainty can be reduced through more data collection, better expert knowledge, or through analytical means.

Reducing aleatory uncertainty through improved process control to lower randomness is application dependent and will not be discussed in this paper other than to note that the identification of the processes that drive uncertainty in the output is invaluable.

This section will discuss improving reported reliability by reducing epistemic uncertainty through targeted testing.

The most direct way to reduce uncertainty in the output (thus reducing 95% CL) is by reducing the variance in the output. Thus, the goal at this stage is to discover which nodes are most responsible for variance in the output. Once that is known, the focus should be on reducing the variance of those nodes. This may involve breaking a single mode into multiple subnodes to increase the level of detail of a particular process.

Saltelli et al. [4] have developed a technique to efficiently determine which variables in a probabilistic model contribute the most to variance in the output. This technique is called variance-based global sensitivity analysis and herein will also be referred to as the Saltelli method.

It is illuminating to compare point or derivative-based sensitivity analysis (previously discussed) and to which Equation 1 refers, with Saltelli global sensitivity analysis.

Conventional derivative (point)-based sensitivities

- Do not take into account uncertainty in the parameters.
- Do provide good information about a parameter at its most likely value.

Global sensitivity analysis (GSA) (the Saltelli method)

- Does take into account uncertainty in the parameters
- Is capable of determining which factors have a major effect on the variance of the POF calculation
- Is capable of determining which factors interact with others in an important way (synergistic effects)
- Is especially useful for determining the small subset of parameters that are important
- Is essentially a variance decomposition algorithm - it determines to some degree what portion of the output variance is due to variance in a particular parameter

The Saltelli process produces two sensitivity measures for each variable. S_i indicates the main effect of variable i, and S_{Ti} indicates the total effect of variable i. There are a few characteristics of these two types of sensitivities that are important to know. S_i indicates by how much one could reduce (on average) the output variance if variable i could be fixed. It is a measure of the main effect. S_{Ti} is useful in determining two important aspects of a variable. This first is if it has interactions with other variables. This can be measured by $(S_{Ti} - S_i)$. The second is if the variable is non-influential and can safely be ignored by setting it to a fixed value when performing time consuming analyses. This is indicated by $S_{Ti} = 0$.

Table 2 shows the results of a Saltelli global sensitivity analysis of the exemplar problem. The variables are sorted from greatest total effect to least total effect. For the example BN configuration, 8 out of 81 nodes have been identified as contributing significantly to variance in the BN output as shown by a total effect of over 0.013. Note that this algorithm tends to 'bottom out' at a non-zero number which in this case is approximately 0.013. Observe that all eight significant nodes are related to inducing or detecting wrinkles in the part.

Reducing epistemic uncertainty using confidence level shifting (CLS)

Now that the nodes causing variance in the output have been identified, the next step is to determine which, how much, and in what order testing should be done to most

Table 2 Saltelli global sensitivity analysis of the exemplar problem

Factor name	Hat_max_load_FE	Hat_max_load_TE	Mode	NumSamples
QA test finds debulking wrinkle if it exists	*0.407223*	*0.633979*	*0.010*	*20*
Wrinkle introduced during debulking	*0.210948*	*0.525119*	*0.010*	*120*
Wrinkle intro. during final cloth overwrap	*0.056164*	*0.187127*	*0.010*	*120*
QA test finds final cloth overwrap wrinkle	*0.033471*	*0.172301*	*0.020*	*120*
QA test finds bagging wrinkle	*0.024144*	*0.154588*	*0.010*	*120*
Wrinkle introduced during release film	*0.029566*	*0.152148*	*0.010*	*120*
QA test finds release film wrinkle	*0.025807*	*0.15214*	*0.010*	*120*
Wrinkle introduced during bagging	*0.033863*	*0.142558*	*0.010*	*120*
Radius thickening intro. during cloth overwrap	0.002573	0.126621	0.020	35
QA test finds debulking radius thickening	0.002591	0.12661	0.010	2

Hat_Max_Load_FE corresponds to S_i whereas Hat_Max_Load_TE corresponds to S_{Ti}. The variables are sorted from greatest total effect to least total effect. For the example model configuration, 8 out of 81 nodes have been identified as contributing significantly to variance in the model output as shown by the italicized entries. Note that all eight nodes are related to inducing or detecting wrinkles in the part. Mode refers to the mode of the beta distribution and NumSamples refers to the parameter 'k' in Equations 4 and 5.

effectively reduce 95% CL. This is known as 'targeted testing' using confidence level shifting (CLS). Confidence levels were explained in Figure 6. Examine Figure 8 in comparison to Figure 6 to understand the idea behind confidence level shifting. To get to Figure 8 from Figure 6, testing would take place to understand if a particular manufacturing step introduces a flaw or not. If not, the step was performed flawlessly and the parameter 'b' should be incremented as described above. A flawless test is also known as a negative result test or NRT. As NRTs accumulate, the beta distribution will narrow and shift to the left. Likewise, its associated 95% confidence level will also shift to the left. This is what is known as confidence level shifting or CLS.

To begin the CLS process, a Monte Carlo procedure will be run for the baseline network and the 95% confidence level of POF will be calculated before any NRTs have been applied to a beta distribution. Note that each beta distribution represents a probability (a factor). Next, a NRT is applied to a single factor and the Monte Carlo analysis is rerun and a new 95% confidence level of POF will be calculated. This provides enough information to calculate a Δ 95% confidence level for POF which is calculated as the original 95% confidence level for POF minus the newly calculated original 95% confidence level for POF. This term can be expressed more compactly as Δ 95% POF or even more simply as ΔPOF. If the cost of performing the test is known, another term, ΔPOF/\$, can be defined which is the amount of change in 95% POF per dollar spent. The metric can be used to determine what data should be collected to most cost effectively drive the 95% POF value to the left.

With the previous information as background and referring to Figure 9, the CLS process can be explained. First determine ΔPOF/\$ for each factor of the set. Add 1 to the 'b' value of the factor with the highest ΔPOF/\$. If the target 95% POF has not been reached continue the process while keeping track of which factors received the 'b' increment and in what order. When the target 95% POF has been reached, the process provides a list of what data should be taken and in what order to most cost-effectively drive the 95% POF value to its target value.

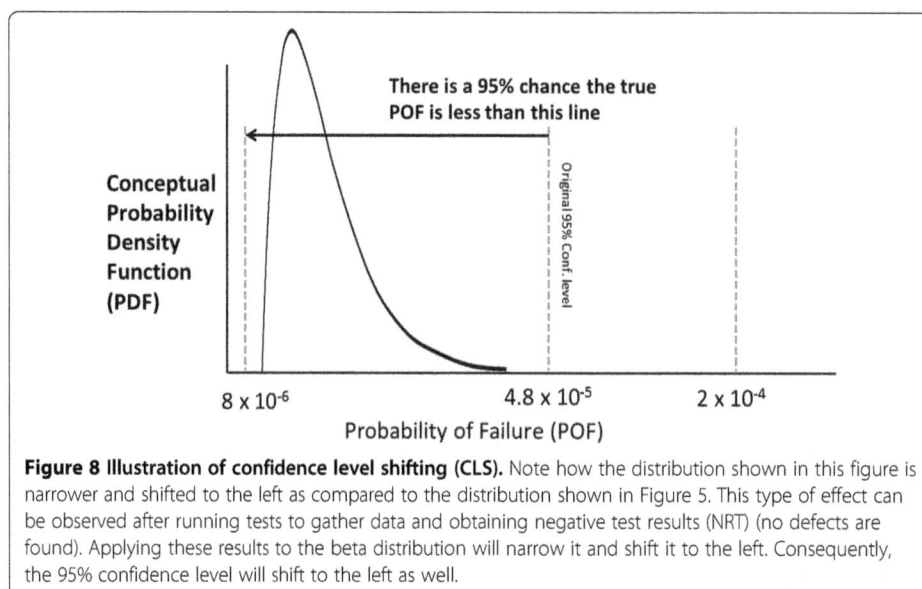

Figure 8 Illustration of confidence level shifting (CLS). Note how the distribution shown in this figure is narrower and shifted to the left as compared to the distribution shown in Figure 5. This type of effect can be observed after running tests to gather data and obtaining negative test results (NRT) (no defects are found). Applying these results to the beta distribution will narrow it and shift it to the left. Consequently, the 95% confidence level will shift to the left as well.

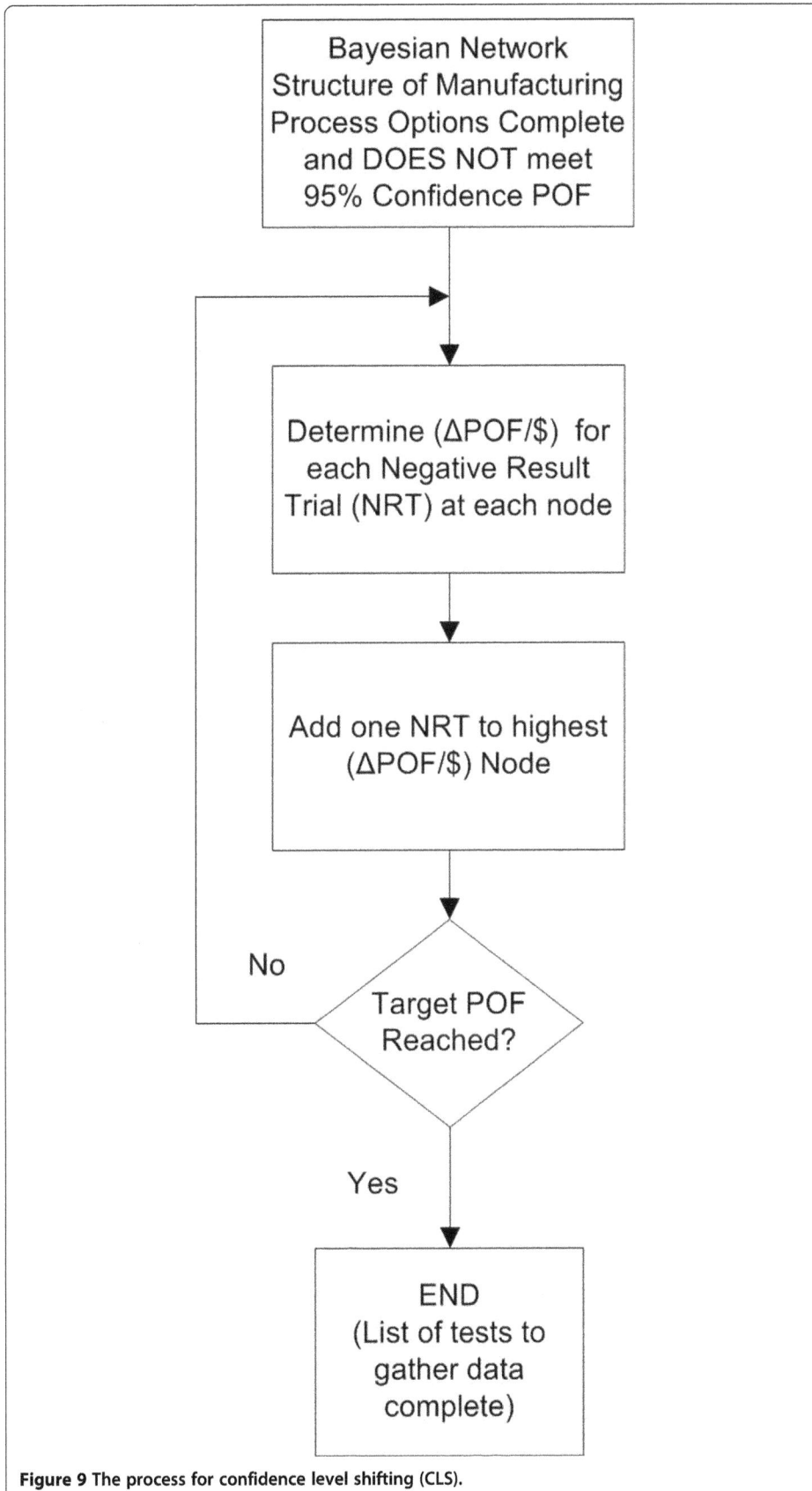

Figure 9 The process for confidence level shifting (CLS).

Note that a single complete fabrication of a part with appropriate inspection will simultaneously provide a data point for every node in the network. Partial part constructions can be accomplished to gather data for just the most important nodes. The ratio of decrease in POF to the cost of running a trial is the measure by which it is decided which trials to run. Note that a test that simultaneously provides data for multiple nodes is called a 'unitized test'. Unitized tests are a time- and cost-efficient technique for generating data to reduce epistemic uncertainty.

Figure 10 shows a plot of decreasing 95% CL POF as a function of (non-unitized) targeted testing. Note that in this example, 700 discrete targeted tests must be run to reduce the 95% CL probability of failure from 4.8×10^{-4} to 1×10^{-4}.

Table 3 is a detailed look at exactly what tests were performed and in what order. Table 3 shows the results of targeted testing analysis using confidence level shifting. Each row of this table provides a breakdown of how much data should be collected for each factor for a maximum reduction in 95% CL POF. For example of 75 collected data points, 25 data points should be collected to check for inducing wrinkles during the final cloth overwrap, and 50 QA tests should be performed to see if wrinkles can be detected during the debulking manufacturing step.

Unitized test for efficient collection of testing data

The example shown in Figure 10 and Table 3 shows the most efficient possible data collection to reduce epistemic uncertainty using individual tests. In this case, however, the burden of testing is high, requiring 700 individual tests to reach a target POF. One technique to reduce this burden is to create a unitized test structure that can test all eight significant features per test. As per Figure 11, only 95 of these unitized tests would have to be performed to reach the target 95% CL POF. This is a sevenfold reduction in the

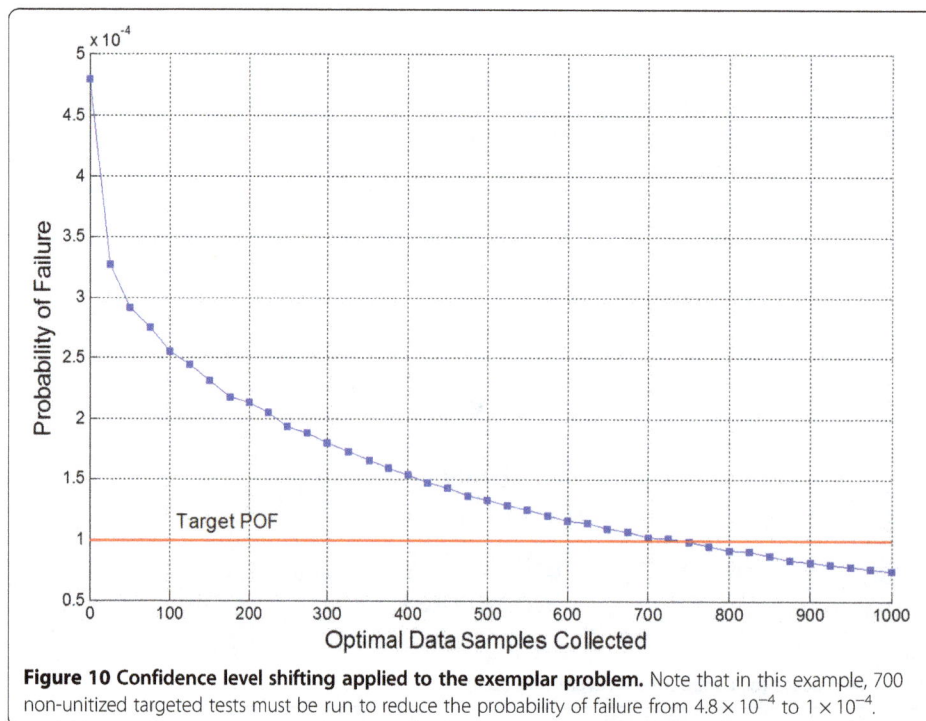

Figure 10 Confidence level shifting applied to the exemplar problem. Note that in this example, 700 non-unitized targeted tests must be run to reduce the probability of failure from 4.8×10^{-4} to 1×10^{-4}.

Table 3 Results of targeted testing analysis using confidence level shifting

Data	MS final cloth overwrap wrinkle	QA final cloth overwrap wrinkle	MS during debulking wrinkle	Q during debulking wrinkle	MS during bagging wrinkle	Q during bagging wrinkle	MS placing release film wrinkle	Q placing release film wrinkle	Lowest POF
0	0	0	0	0	0	0	0	0	4.80×10^{-4}
25	0	0	0	25	0	0	0	0	3.28×10^{-4}
50	0	0	0	50	0	0	0	0	2.92×10^{-4}
75	0	25	0	50	0	0	0	0	$2.75 \times 10 - 4$
100	0	25	0	75	0	0	0	0	2.56×10^{-4}
125	25	25	0	75	0	0	0	0	2.45×10^{-4}
150	25	25	0	75	0	25	0	0	2.31×10^{-4}
175	25	25	0	75	0	25	25	0	2.18×10^{-4}
200	25	25	0	75	0	25	25	25	2.13×10^{-4}
225	25	50	0	75	0	25	25	25	2.05×10^{-4}
250	25	50	0	100	0	25	25	25	1.94×10^{-4}
275	25	50	0	100	0	25	25	50	1.88×10^{-4}
300	25	75	0	100	0	25	25	50	1.80×10^{-4}

Each row of this table provides a breakdown of how much data should be collected for each factor for a maximum reduction in 95% CL POF. For the example of 75 collected data points (shown in italics), 25 data points should be collected to check for inducing wrinkles during the final cloth overwrap, and 50 QA tests should be performed to see if wrinkles can be detected during the debulking manufacturing step.

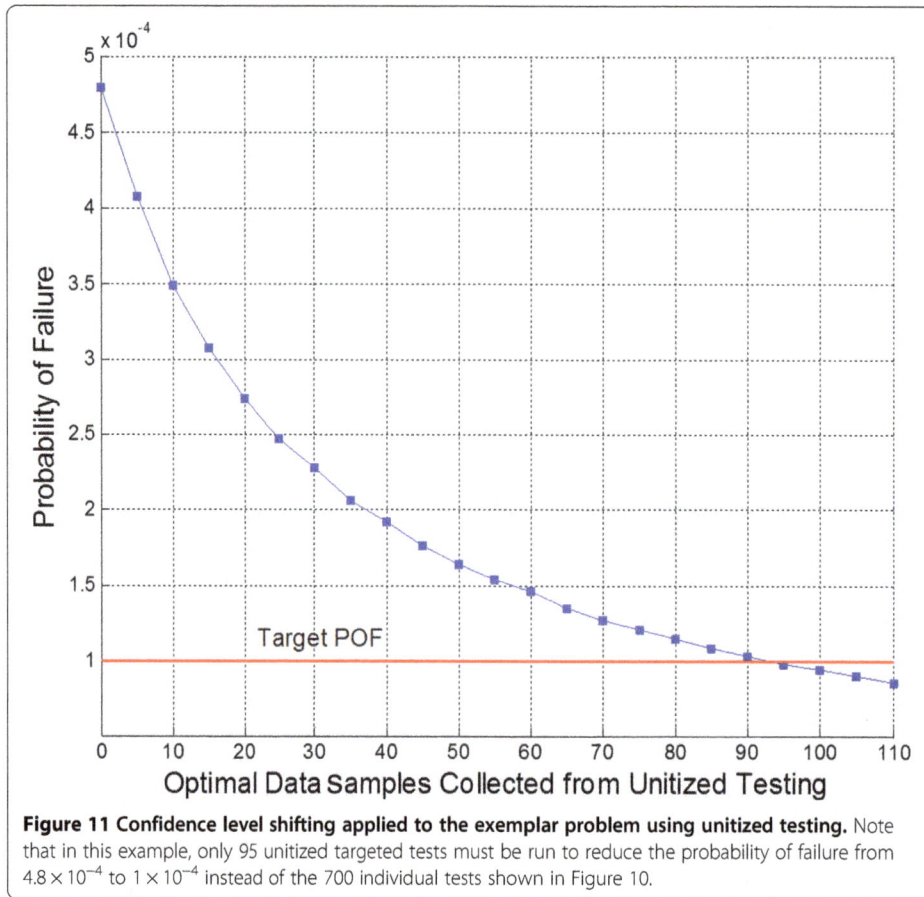

Figure 11 Confidence level shifting applied to the exemplar problem using unitized testing. Note that in this example, only 95 unitized targeted tests must be run to reduce the probability of failure from 4.8×10^{-4} to 1×10^{-4} instead of the 700 individual tests shown in Figure 10.

number of tests. Despite this large reduction in the number of required tests, in some cases, this will still be too expensive or time-consuming. A few points should be noted however. The first is that every problem will be different. For example, in some cases, the number of tests may be reduced from 70 down to 10. In addition, CLS is only one of the tools used to improve the 95% CL. As discussed above, reducing aleatory variability through the use of QA tests and better process control are alternative options that can be used in addition to or in place of CLS depending in the problem at hand.

Results and discussion

Putting it all together - an example using credibility tools

This section will provide an example of using the credibility tools discussed in this paper to reach a 95% confidence level probability of failure of 1×10^{-4} when starting with a manufacturing process for a three-hat-stiffened panel that has a 4.4% probability of failure under certain environmental conditions when no quality assurance testing is done.

As discussed in detail in reference [3], a Bayesian network is constructed for the three-hat-stiffened panel that includes all possible manufacturing options, including many potential quality assurance (QA) tests to catch defects both during the manufacturing process and as a final check. Using the network's point probabilities, it is possible to quickly evaluate all possible combinations of options to find the highest reliability part at any given price point or conversely the lowest probability of failure. Figure 12 is plot of optimal POF for any given price point. A few conclusions can be

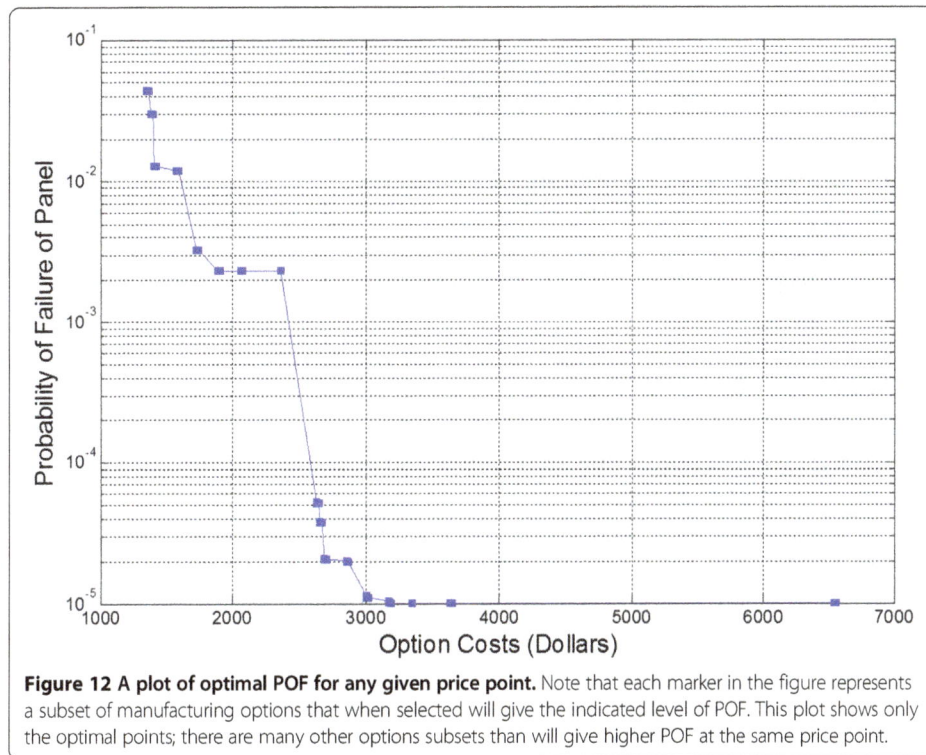

Figure 12 A plot of optimal POF for any given price point. Note that each marker in the figure represents a subset of manufacturing options that when selected will give the indicated level of POF. This plot shows only the optimal points; there are many other options subsets than will give higher POF at the same price point.

drawn from Figure 12. This first is that the manufacturing process is highly dependent on QA checks for reliability. With no quality checks, the POF is .044 or 4.4% per part (the leftmost marker in the plot). The POF can range as low as 1.01×10^{-5} with a full set of QA checks in place. It should be noted that this plot does not include the price of rework or scrap due to faulty manufacturing. Future work will address this issue.

To reach the stated goal of 1×10^{-4} POF or 0.9999 reliability, the \$2,626 option is the most cost-effective (not including scrap or rework). This option represents the case that all QA tests are off except for the ultrasonic inspection for wrinkle QA test.

The Bayesian network with this configuration is evaluated using Monte Carlo analysis to include the effects of uncertainty in the expert elicited opinions. The result is shown in Figure 13.

While the mode of this analysis meets the goal of 1×10^{-4} (being 5.6×10^{-5}), the 95% confidence value in POF is slightly too large at 2.3×10^{-4}. Another issue with using only the ultrasonic inspection for wrinkle QA test is that it only catches the wrinkle after the part is complete, leading to a very high rejection rate of a finished part. According to the model, there is a 46% chance of a wrinkle defect. This means that nearly half of the completed parts would have to be rejected. This is unacceptable. To better understand what is causing the wrinkles, an analysis of manufacturing steps as modeled by the Bayesian network is undertaken.

By removing all QA tests and running a Saltelli global sensitivity analysis, the manufacturing steps most responsible for output variation can be found.

Looking at Table 4, it is clear that wrinkles are the primary cause of increased failure and that there are four manufacturing steps that contribute to wrinkles. Given the high mode values (these are used as the point probabilities) it is also clear that the probability of incurring wrinkles during the manufacturing process is quite high.

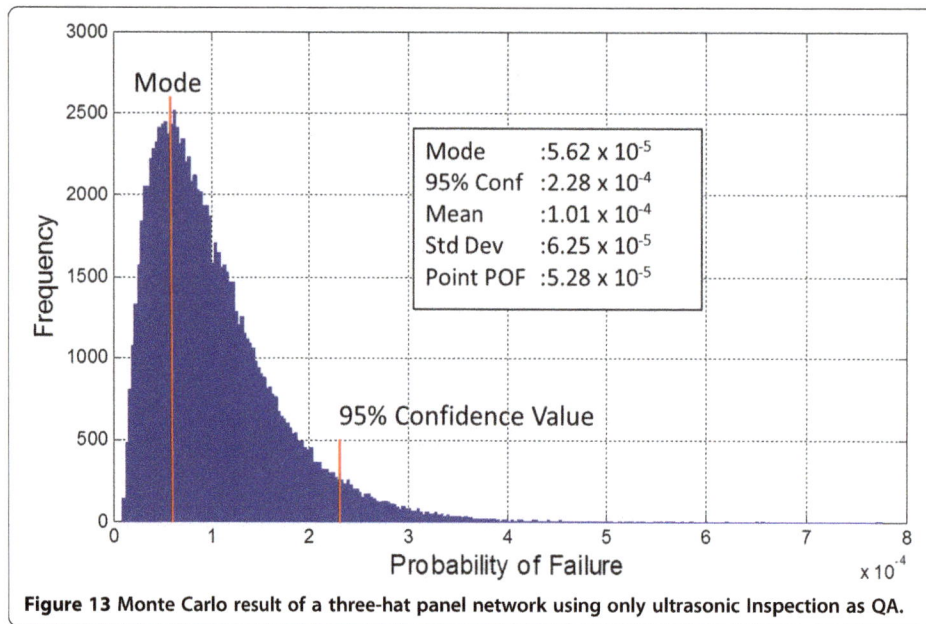

Figure 13 Monte Carlo result of a three-hat panel network using only ultrasonic Inspection as QA.

The relationship between p (wrinkle) and POF

Now that it is clear the wrinkles are the primary driver behind increased POF for this particular example, it is informative to note the relationship between the two quantities. Figure 14 shows a plot of POF vs. p (wrinkle). Note that by Figure 14, POF is linearly related to p (wrinkle) and POF is generally 0.096 of p (wrinkle). Thus, to get a POF of 1×10^{-4} or better, p (wrinkle) must be 1×10^{-3}. Occasionally, it is simpler to use p (wrinkle) as a proxy for POF, as this is directly displayed in the Bayesian network rather than needing to be calculated outside of it.

Examining the effect of four in-process QA checks

Observing the mode values of Table 4, the major issue in reaching a POF of 1×10^{-4} is not the uncertainty around the manufacturing process but the high mode value. The high mode value indicates that the manufacturing steps have a very high probability of inducing a wrinkle. Improving the actual manufacturing process reduces or eliminates rework or scrap and for that reason is normally considered the best way to improve quality and will be considered in the next section. The most expeditious way, however, to attack this problem without expending effort improving the manufacturing process is to apply quality assurance tests (QA tests) at the four manufacturing steps that can induce wrinkling in order to find and correct the wrinkles at the stage where they are introduced. The four tests, their costs, and their relative effectiveness are shown in Table 5. These tests are included in the Bayesian model to view their effect in p (wrinkle).

Table 4 Saltelli global sensitivity analysis results

Title	Hat_max_load_FE	Hat_max_load_TE	Mode	NumSamples
Wrinkle placing release film strips	0.435196	1.101102	0.2153	120
Wrinkle induced during debulking	0.192996	0.950746	0.1136	120
Wrinkle during bagging and pleating	0.713702	0.824299	0.2153	120
Wrinkle applying final cloth over wrap	0.039608	0.776457	0.0100	120

Top four variables contributing to variance identified. Hat_Max_Load_FE corresponds to S_i whereas Hat_Max_Load_TE corresponds to S_{Ti}. The variables are sorted from greatest total effect to least total effect. For the example model configuration, 4 out of 81 nodes have been identified as contributing significantly to variance in the model output. Note that all four nodes are related to inducing or detecting wrinkles in the part. Mode refers to the mode of the beta distribution and NumSamples refers to the parameter 'k' (expert confidence) in Equations 4 and 5.

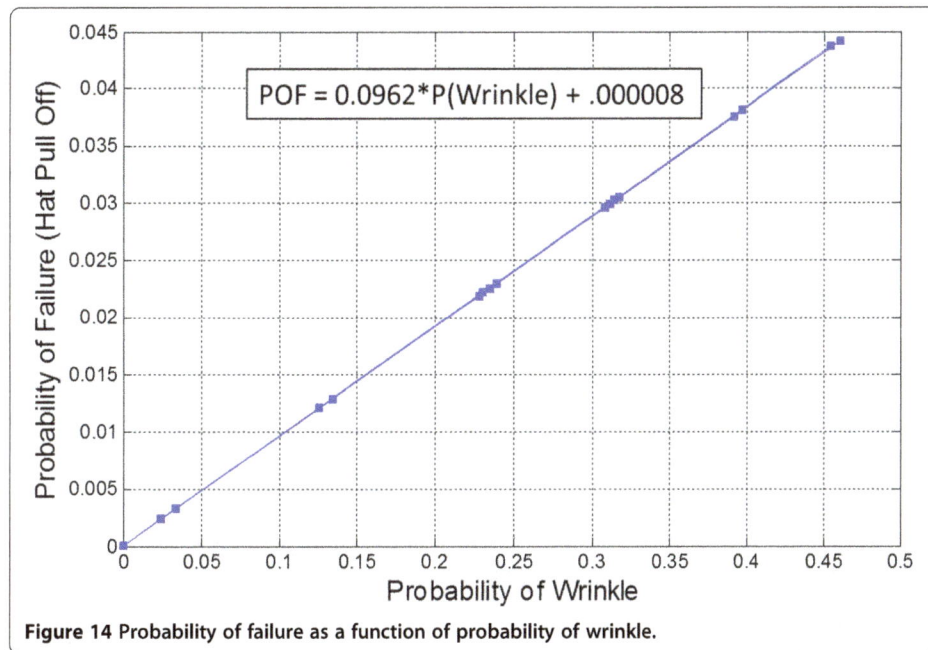

Figure 14 Probability of failure as a function of probability of wrinkle.

As Figure 15 shows, the p (wrinkles) is reduced from 46% to 2.43% by applying all four of these QA tests. Although that is a significant drop, p (wrinkle) must be lowered to 0.1% (1×10^{-3}) in order to reach the reliability target.

Identifying manufacturing process areas to target for improvements

Based on the point probability of a wrinkle being 0.0243 and needing to be 0.001, it is clear that process improvements must be made. Both global and point sensitivity analysis are performed on the four QA test model to determine the focus areas. Tables 6 and 7 show the results of this analysis.

Table 6 shows that most of the variance is due to the debulking and bagging steps and also that there is synergy between those nodes and other nodes. The point sensitivity analysis of Table 7 shows that those nodes are also prominent in affecting POF. Based on these two tables, it appears that the most efficient means of decreasing POF is by improving the manufacturing debulking related steps and the QA of the debulking simultaneously (top two globally sensitive) to take advantage of the synergy between them as well as their high point sensitivity. If this is not enough, then the nodes related to bagging and the release film should be worked on next.

To simulate working on the debulking process, it is assumed for the purposes of this example that engineers improve the process by examining and improving such elements as the bulk factor of the product form, tack or lack of tack, and the debulk process itself.

Table 5 Efficacy and cost of QA tests for wrinkles

Node	Mode	NumSamples	Cost
QA test finds bagging wrinkle	0.0400	120	$30
QA test finds debulking wrinkle	0.1000	2	$315
QA test finds release film wrinkle	0.0200	120	$30
QA test finds final cloth overwrap wrinkle	0.0200	120	$165

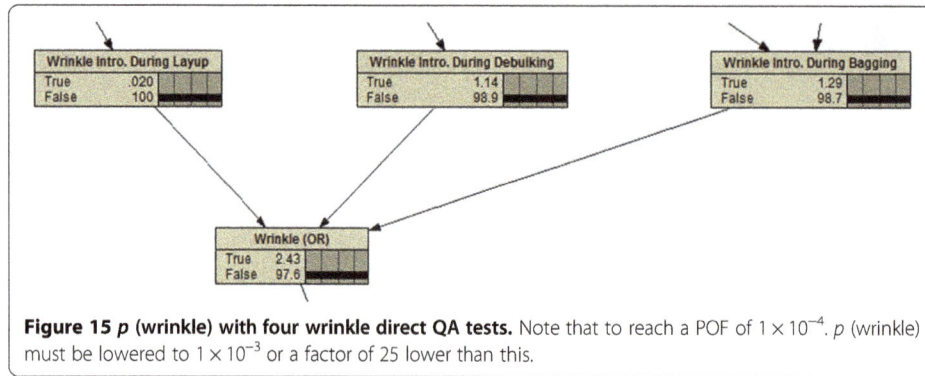

Figure 15 p (wrinkle) with four wrinkle direct QA tests. Note that to reach a POF of 1×10^{-4}. p (wrinkle) must be lowered to 1×10^{-3} or a factor of 25 lower than this.

After those improvements, there is only a .01 chance of inducing a wrinkle during the debulking process (improved from .11) and a .99 chance of finding a wrinkle at that point with a 20 sample size. With these improvements in place, the point probability network results are shown in Figure 16. This figure shows that the probability of wrinkle is still much higher than the 0.1% needed and that the major source of the high probability appears to be due to the steps making up bagging.

To help verify this conclusion, another global sensitivity analysis is performed with the results shown in Table 8. This table shows that the top four nodes contributing to output variance are now all related to bagging. Note that placing release film is part of the bagging process. Since Table 8 verifies what was seen in Figure 16, it is clear that improving the nodes related to inducing wrinkles during the bagging process would most benefit POF. This time, it is assumed that engineers improve the bagging process steps such that there is a 0.01 chance of wrinkle and improve the QA tests such that there is a 0.99 chance of detection.

As shown in Figure 17, these changes improve the point model probability of wrinkle to 0.05% which is better than the goal of 0.1%.

At this point, a Monte Carlo analysis must be run to take into account the uncertainty in the model parameters, and a 95% confidence level must be calculated to add the necessary amount of conservatism to the estimate. Figure 18 shows the results of the Monte Carlo process. Note that the mode of the POF distribution at 1.2×10^{-4} is close to the target value of 1×10^{-4}, but the 95% confidence level is 4.93×10^{-4}, which is about a factor of five from the target level.

At this stage, it may be cost-effective to perform targeted testing to reduce epistemic uncertainty by using the confidence level shifting (CLS) analysis technique. The first

Table 6 Global sensitivity analysis of the manufacturing network that includes the four wrinkle direct QA tests

Factor name	Hat_max_load_FE	Hat_max_load_TE	Mode	NumSamples
QA test finds debulking wrinkle if it exists	*0.817142*	*0.88832*	*0.1000*	*2*
Wrinkle introduced during debulking	*0.13929*	*0.279974*	*0.1136*	*120*
QA test finds bagging wrinkle	*0.003301*	*0.096139*	*0.0400*	*120*
QA test finds release film wrinkle	0.00201	0.087771	0.0200	120
Wrinkle introduced during bagging	0.000163	0.080021	0.2153	120
Wrinkle introduced during release film	−0.003777	0.082574	0.2153	120
QA test finds final cloth overwrap wrinkle	−0.005413	0.082446	0.0200	120
Wrinkle intro. during final cloth overwrap	−0.005268	0.082371	0.0100	120

The italicized entries indicate the significant variables identified through the sensitivity analysis.

Table 7 Point sensitivity analysis of the manufacturing network that includes the four wrinkle direct QA tests

Node	POFHat_max_load_sens	Mode	NumSamples
QA test finds bagging wrinkle	*0.02038*	*0.040*	*120*
Wrinkle introduced during release film	*0.02029*	*0.020*	*120*
QA test finds debulking wrinkle if t exists	*0.01079*	*0.100*	*2*
Wrinkle introduced during debulking	*0.00949*	*0.114*	*120*
Wrinkle introduced during bagging	*0.00379*	*0.215*	*120*
Wrinkle introduced during release film	0.00189	0.215	120
Wrinkle intro. during final cloth overwrap	0.00188	0.010	120
QA test finds final cloth overwrap wrinkle	0.00094	0.020	120

The italicized entries indicate the significant variables identified through the sensitivity analysis.

step in the CLS process is to identify which nodes are causing the most variance on the output POF using global sensitivity analysis. Table 2 from the main body of this paper shows the results of this. With the steps and QA tests related to wrinkles balanced in terms of performance by engineers, all eight nodes related to wrinkles are found to be important. Figure 9 from this paper's CLS section shows that 95 successful unitized targeted tests can be run to drive the 95% CL POF to 1×10^{-4}.

Now that the manufacturing process has been modified and additional testing performed to drive the 95% CL POF down to 1×10^{-4}, the final step is to perform a sensitivity analysis of the 95% confidence level POF to a change in mode, which is a good indication of influential distributions that must be justified by documentation. Table 9 shows the results of this analysis. The table is sorted by node with those nodes that have the most influence starting at the top. The top eight influential nodes turn out to be the eight nodes that had 95 unitized tests performed to make their estimates more certain. Due to the collection of this extra confirmatory data, these nodes can be considered credible. The ninth node in the list - radial thickening (RT) during debulking (not shown in Figure 3) has a sensitivity of 0.28×10^{-5} per 0.01 of mode change. This means that the given mode (0.01) could be as high as .04, or four times higher than the given mode, before the 95% CL POF reaches the target value. This value is thus judged to be credible.

Summary of using credibility tools to reach a credible target 95% CL POF

In summary, an example project consisting of manufacturing a three-hat-stiffened panel was used as a case study for exercising the credibility tools detailed in this paper. The goal of the example was to analyze a manufacturing process in terms of the factors that

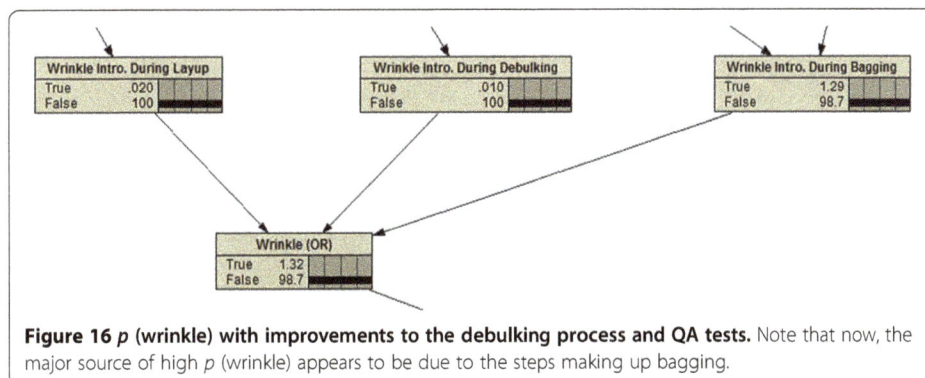

Figure 16 p (wrinkle) with improvements to the debulking process and QA tests. Note that now, the major source of high p (wrinkle) appears to be due to the steps making up bagging.

Table 8 Global sensitivity analysis of the manufacturing network that includes four wrinkle direct QA tests and improved debulking steps

Factor name	Hat_max_load_FE	Hat_max_load_TE	Mode	NumSamples
QA test finds bagging wrinkle	*0.488152*	*0.625328*	*0.0400*	*120*
QA test finds release film wrinkle	*0.298514*	*0.388214*	*0.0200*	*120*
Wrinkle introduced during bagging	*0.135016*	*0.145427*	*0.2153*	*120*
Wrinkle introduced during release film	*0.039348*	*0.123047*	*0.2153*	*120*
Wrinkle introduced during debulking	0.00898	0.122452	0.0100	120
QA test finds debulking wrinkle if it exists	0.036094	0.115424	0.0100	20
Wrinkle intro. during final cloth overwrap	0.006912	0.096334	0.0100	120
QA test finds final cloth overwrap wrinkle	0.003504	0.095578	0.0200	120
Radius thickening intro. during final cloth overwarp	0.003001	0.093529	0.0200	35

This table shows that the top four nodes (in italics) contributing to output variance are now all related to bagging. Note that placing release film is part of the bagging process.

contribute to its unreliability, to ensure that the expert opinion using to furnish the parameters of that model were credible, and then use a number of tools to determine the optimal way to create a more reliable product that met a target reliability number. Note that this was accomplished conceptually for illustrative purposes.

The examination started with noting the effect of any and all possible combinations of manufacturing options on the point reliability of the part. It was found that quality assurance tests had an extremely high impact on part reliability. After noting that a post-manufacturing QA test was effective but that it resulted in many costly part rejections, another analysis was undertaken which found four manufacturing steps that induced wrinkles also contributed the most to variance in the output. QA tests were directly applied within the model to address these four steps but it was found that they were not effective enough to reach the target 95% CL POF. An iterative process was then undertaken which involved identifying and then improving aspects of the manufacturing process until the process nearly reached the target POF. At this point, an effort to reduce epistemic uncertainty in the model was undertaken using confidence level shifting to identify target testing. This testing optimally reduced uncertainty in the model. Finally, a sensitivity analysis of the 95% confidence level POF to a change in mode was performed for each node of the model to indicate influential distributions that must be justified by

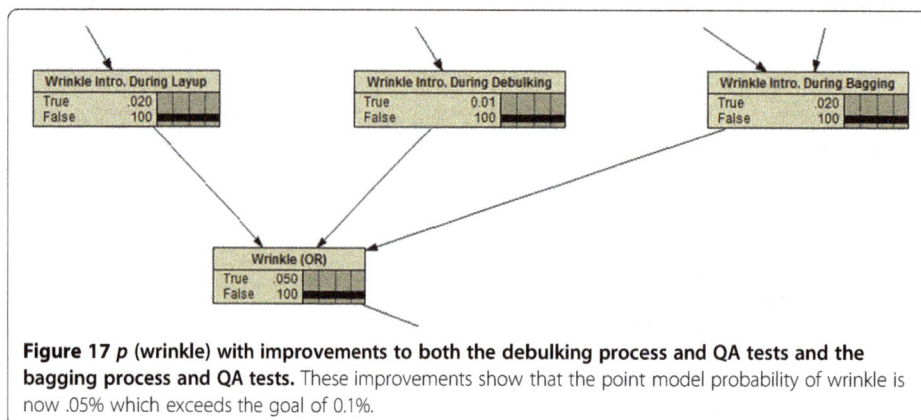

Figure 17 *p* **(wrinkle) with improvements to both the debulking process and QA tests and the bagging process and QA tests.** These improvements show that the point model probability of wrinkle is now .05% which exceeds the goal of 0.1%.

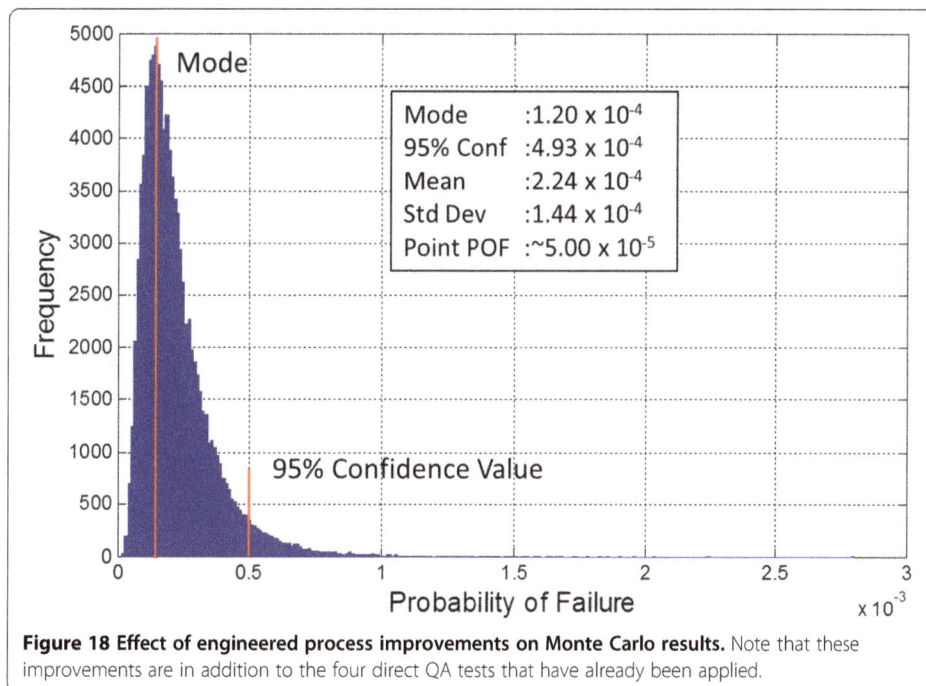

Figure 18 Effect of engineered process improvements on Monte Carlo results. Note that these improvements are in addition to the four direct QA tests that have already been applied.

documentation. At this point, it was found that with the targeted testing that had already occurred that the model was credible.

Conclusions

This paper details an approach to obtaining credible model output when model parameters are based on expert opinion. Although the model used as an example in this paper is a Bayesian network model, the approach and techniques described in this paper are completely transferrable to any model using uncertain parameters. This paper details an approach to obtaining credible model output based on the idea of having a hypothetical expert whose unconscious bias influences the model output and discovering and using countermeasures to find and prevent these biases. Countermeasures include replacing point probabilities with beta distributions to incorporate uncertainty and requiring 95% confidence levels to add conservatism. Multiple types of sensitivity analyses are used to identify parameters in the model that have the most influence over the model's output. This includes a derivative point probability-based sensitivity analysis that is a good indicator of relevance when all parameters are at their most likely values, a sensitivity analysis of 95% confidence level to a change in mode which is a good indication of influential distributions that must be justified by documentation and a variance-based global sensitivity analysis which is useful for identifying which model parameters contribute the most to output variance and which model parameters have synergy with other model parameters. Finally, this paper uses a new technique named 'confidence level shifting' to cost and time optimally reduce epistemic uncertainty in the model. This is useful when uncertainty in model parameters is inflating the 95% confidence level of a reported target output (such as probability of failure or probability of a defect) and needs to be brought down as cost effectively as possible.

Table 9 Results of sensitivity calculations of 95% CL to a change in mode

Nodes	numSamples	Old_mode	New_mode	Delta_mode	95POF	New95POF	Delta_POF	Sen
Wrinkle introduced during debulking	215	0.0055	0.0155	0.0100	9.26×10^{-5}	1.21×10^{-4}	2.81×10^{-5}	2.81×10^{-3}
QA test finds debulking wrinkle if it exists	115	0.0055	0.0155	0.0100	9.26×10^{-5}	1.14×10^{-4}	2.11×10^{-5}	$2.11 \times 10 - 3$
QA test finds bagging wrinkle	215	0.0055	0.0155	0.0100	9.26×10^{-5}	1.11×10^{-4}	1.85×10^{-5}	1.85×10^{-3}
QA test finds final cloth overwrap wrinkle	215	0.0032	0.0132	0.0100	9.26×10^{-5}	1.11×10^{-4}	1.81×10^{-5}	1.81×10^{-3}
Wrinkle introduced during bagging	215	0.0055	0.0155	0.0100	9.26×10^{-5}	1.10×10^{-4}	1.75×10^{-5}	1.75×10^{-3}
QA test finds release film wrinkle	215	0.0055	0.0155	0.0100	9.26×10^{-5}	1.10×10^{-4}	1.75×10^{-5}	1.75×10^{-3}
Wrinkle introduced during release film	215	0.0055	0.0155	0.0100	9.26×10^{-5}	1.09×10^{-4}	1.60×10^{-5}	1.60×10^{-3}
Wrinkle intro. during final cloth overwrap	215	0.0055	0.0155	0.0100	9.26×10^{-5}	1.06×10^{-4}	1.37×10^{-5}	1.37×10^{-3}
Radius thickening intro. during debulking	30	0.0100	0.0200	0.0100	9.26×10^{-5}	9.55×10^{-5}	2.88×10^{-6}	2.88×10^{-4}
QA test finds release film up. radius thickening	10	0.0100	0.0200	0.0100	9.26×10^{-5}	9.51×10^{-5}	2.50×10^{-6}	2.50×10^{-4}
QA test finds bagging radius thickening	20	0.0050	0.0150	0.0100	9.26×10^{-5}	9.45×10^{-5}	1.93×10^{-6}	1.93×10^{-4}

The top eight nodes are italicized. The sensitivity of the mode of for an average one of these nodes is roughly 2×10^{-3} or 2×10^{-5} for every .01 change in mode. This effectively means that if the expert started with a mode of .008 and reduced it down to 0.001, a magnitude change in 95% CL POF would have occurred.

Abbreviations

Δ: delta, change in value; A: beta distribution parameter expressing the number of flawed examples; B: beta distribution parameter expressing the number of flawless examples; BN: Bayesian networks; CL: confidence level; CLS: confidence level shifting; CMTI: Certification Methodology to Transition Innovation; DARPA/DSO: Defense Advanced Research Projects Agency Defense Science Office; GSA: global sensitivity analysis; K: expert confidence in estimate in terms of equivalent prior sample size; Mode: the most likely probability of a flaw; NRT: negative result test; OM: Open Manufacturing; P: proportion of flaws in the beta distribution; P_i: probability of node i inducing or failing to detect a defect; POF: probability of failure; QA: quality assurance; RT: radial thickening; $S_{95\%CL}$: sensitivity of the 95% CL of model output due to change in mode of node i; S_{Di}: derivative-based sensitivity measure; Si: effect due to variable i; S_{Ti}: total effect due to variable i; X_i: model parameter; Y: model output.

Competing interests

The author declares that he has no competing interests.

Acknowledgements

This paper is sponsored by Defense Advanced Research Projects Agency, Defense Sciences Office under the Open Manufacturing Program, ARPA Order No. S587/00, Program Code 2D10, issued by DARPA/CMO under contract no. HR 0011-12-C-0034. The views and conclusions contained in this document are those of the authors and should not be interpreted as representing the official policies, either expressly or implied, of the Defense Advanced Research Projects Agency of the U.S. Government. This paper was approved for public release, distribution unlimited as 14-00070-EOT.

References

1. Renieri G (2013) High performance stiffened panel design concept. Annual Report. Office of Naval Research (ONR), Arlington, VA. Contract N00014-12-C0417
2. Koller D, Friedman N (2009) Probabilistic graphical models: principles and techniques. Press, MIT, Cambridge
3. Hahn GL, Pado LE, Thomas MJ, Liguore SL (2014) Application of risk quantification approach to aerospace manufacturing using Bayesian networks. Paper presented at AIAA SciTech 2014. National Harbor, Maryland, pp 13–17. Report AIAA-2014-0463
4. Saltelli A, Ratto M, Andres T, Campolongo F, Cariboni J, Gatelli D, Saisana M, Tarantola S (2008) Global sensitivity analysis. The Primer, Wiley, John & Sons, Incorporated.
5. Kruschke J (2011) Doing Bayesian data analysis. Press, Academic, New York
6. Kruschke J (2012) Beta distribution parameterized by mode instead of mean. http://doingbayesiandataanalysis. blogspot.com/2012/06/beta-distribution-parameterized-by-mode.html
7. Haldar A, Mahadevan S (2000) Probability, reliability, and statistical methods in engineering design. Wiley, John & Sons, Incorporated, New York
8. Milton JS, Arnold JC (1986) Probability and statistics in the engineering and computing sciences. Elsevier Academic Press, Burlington, MA
9. Abrahamson N (2006) Seismic hazard assessment: problems with current practice and future developments Proceedings. First European Conference on Earthquake Engineering and Seismology, Geneva, Switzerland. September 2006

9

Combining material and model pedigree is foundational to making ICME a reality

Steven M Arnold[*], Frederic A Holland Jr, Brett A Bednarcyk and Evan J Pineda

* Correspondence:
Steven.M.Arnold@nasa.gov
NASA Glenn Research Center, 21000
Brookpark Road, Cleveland, OH
44135, USA

Abstract

With the increased emphasis on reducing the cost and time to market of new materials, the need for analytical tools that enable the virtual design and optimization of materials throughout their processing-internal structure-property-performance envelope, along with the capturing and storing of the associated material and model information across its life cycle, has become critical. This need is also fueled by the demands for higher efficiency in material testing; consistency, quality, and traceability of data; product design; engineering analysis; as well as control of access to proprietary or sensitive information. Fortunately, materials information management systems and physics-based multiscale modeling methods have kept pace with the growing user demands. Herein, recent efforts to identify best practices associated with these user demands and key principles for the development of a robust materials information management system will be discussed. The goals are to enable the connections at various length scales to be made between experimental data and corresponding multiscale modeling toolsets and, ultimately, to enable ICME to become a reality. In particular, the NASA Glenn Research Center efforts towards establishing such a database (for combining material and model pedigree) associated with both monolithic and composite materials as well as a multiscale, micromechanics-based analysis toolset for such materials will be discussed.

Keywords: Information Management; Informatics; Data schema; Analysis; Experimental Data; Simulation Data; Pedigree; Multiscale Modeling; Micromechanics

Background

With the increased emphasis on reducing the cost and time to market of new materials, ICME (Integrated Computational Materials Engineering) has become a fast-growing discipline within materials science and engineering. The vision of ICME is compelling in many respects, not only for the value added in reducing time to market for new products with advanced, tailored materials but also for enhanced efficiency and performance of these materials. Although the challenges and barriers (both technical and cultural) are formidable, substantial cost, schedule, and technical benefits can result from broad development, implementation, and validation of ICME principles [1]. ICME is an *integrated* approach to the design of products, and the materials that comprise them, by linking material models at multiple time and length scales.

A key ingredient is the linkage with manufacturing processes, which produce internal material structures, and in turn influence material properties and allowables, enabling tailoring (engineering) of materials to specific industrial applications. Figure 1 illustrates

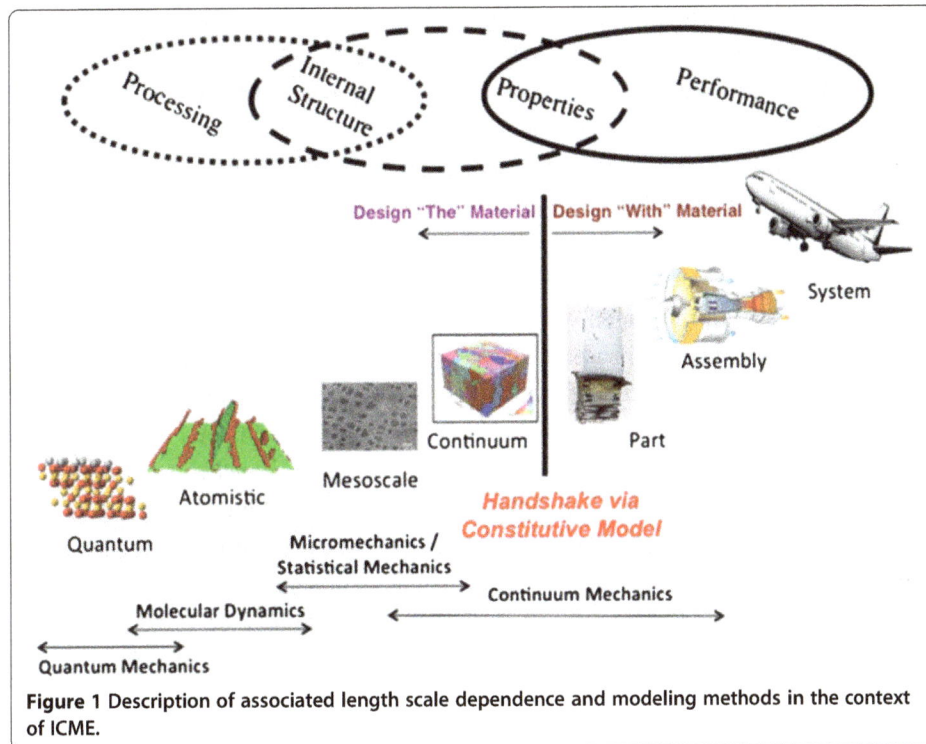

Figure 1 Description of associated length scale dependence and modeling methods in the context of ICME.

the interconnection of these scales and their cause/effect relationships, e.g., processing conditions produce a particular microstructure from which properties are obtained, which then dictate a specific structural performance. Note that the evolution of elliptical line types (i.e., dotted to dashed to solid line) are purposely included to imply the level of maturity/understanding (from immature, to semi-mature, to mature, respectively) of modeling at each level of scale (both temporal and geometric). Furthermore, the figure illustrates the difference between two non-exclusive viewpoints, that is, designing 'with-the-material' (structural analyst viewpoint) versus designing 'the material' (a materials scientist viewpoint). It is also apparent that the fundamental linkage between these two viewpoints is ultimately the associated constitutive model(s) for a particular material. One cannot overestimate the importance of understanding the input and output at each scale in order to determine the appropriate 'handshaking' between scales and the meaningful properties that are ultimately required by a structural analyst.

Equally important is the fact that experiments (whether computational/virtual or laboratory) performed at a given level can be viewed from two perspectives. If one 'looks up' to higher scales, then the results can be viewed as exploration or characterization experiments used to identify/obtain the necessary model features or parameters, respectively, operating at the present and/or next higher level. Conversely, if one 'looks down', these same results can be used to validate the modeling methods/approaches employed to transition from the lower level(s) to the given level.

While there is a clear indication that ICME is growing, utilization of ICME in the daily work of researchers and engineers is still lacking. The key contributing factors, since ICME is an inherently data-intensive activity, are the lack of a robust information management system and the lack of a digital storage culture within most organizations. This stems from the fact that on the surface, a materials properties database may seem simply like a fancy means of storing, retrieving, and distributing materials data, something akin to an electronic file cabinet. However, as discussed by Marsden et al. [2] and Arnold et al. [3], an effective ICME materials database (e.g., one in which experimental and computational mechanics are fully coupled) must allow the data inside a database to be seamlessly accessible by analysis tools and allow the results from analyses to be read back into the database and stored with all of the associated metadata, while keeping track of associations across the full range of length scales.

For example, for a physics-based model to predict the yield strength of a nickel-based superalloy it may need to draw upon quantum mechanics predictions of stacking fault energies, lattice distortions, and phase equilibria of several different alloying elements. These predictions would be combined with microstructural scale models that either use the quantum mechanics predictions or are calibrated with experimental data. Phase equilibria models such as CALPHAD° models are an example, as well as processing-microstructure models of castings or forgings. Important information necessary for a yield strength model would include not only equilibrium phases but also the kinetics of microstructural evolution (of several features, including γ' precipitate and carbide size and spacing, grain size and grain boundary phases). The maturity of these models already allows semi-quantitative predictions of various parameters, but the development of higher fidelity models will require the capture, analysis, and dissemination of higher fidelity data, as well as all associated pedigree information for calibration and validation. For example, while a current model may utilize an average particle size as a key parameter, future models may require the entire particle size and shape distributions to be measured and tracked with respect to various manufacturing methods. Clearly, the enormity of data types (e.g., discrete, functional, structured, and unstructured) and the sheer quantity of data can be overwhelming. Consequently, historical static data systems are likely to be gradually phased out, evolving to become an integral part of dynamic materials property databases that are web-accessible and in which data - and the relationships between items of data - can be interactively searched, reorganized, analyzed, and applied. These dynamic databases have great superiorities in satisfying the needs of modern materials-related sciences and engineering focused activities like ICME.

Furthermore, it is critical to understand that ICME is not just developing processing-microstructure (P-M) relationships or microstructure-property (M-P) relationships independently, rather it is the full integration of these various length scale-specific relationships, wherein linkages from processing all the way up to performance can be made and utilized. This requirement greatly increases the need for data/metadata and contextual linkage so that knowledge can be both captured and discovered. For example, the variety and complexity of modern materials, and their applications, necessitate complicated, and often extensive, materials testing. As for composite materials, large volumes of test data on various forms of the composites themselves, as well as individual constituents' thermal and mechanical behavior, are often required. Given a micromechanics-based analysis approach, it is typical to require that data for each constituent be reliably and conveniently traced back from the final products through their processing steps to the original raw materials. A second example is the need to provide adequate data to support increasingly sophisticated nonlinear, anisotropic, and multi-scale engineering analyses. Here again, instead of storing a simple set of reduced, point-wise data, like elastic modulus and yield strength, the entire response (e.g., stress-strain, creep, and relaxation) curve may be required. Collating, storing, processing, interacting with, and finally applying such data and metadata require advanced dynamic information systems, enabling management of changing proprietary data alongside reference data collections, while ensuring consistency, quality, applicability, and traceability.

Prior publications [3-6] discussed the data scheme, best practices, and informatics required to establish a robust, twenty-first century information management system for capturing and analyzing materials information. The goal of the information management system is to enable 1) generalized constitutive modeling and 2) data mining to establish microstructure/property/failure relationships for monolithic and composite materials. The proposed schema/requirements for ICME were demonstrated using a turbine disk Ni-based superalloy, in Arnold et al. [3]. Furthermore, Arnold et al. [6] argued that integrating both virtual (computationally based) and experimental data, over the entire material data life cycle (see Figure 2) and at various length scales, in the same information management system is essential for ICME to become a reality and to permeate the material and engineering cultures within a given organization. For example, Figure 3 illustrates the interaction between experimental data and virtual data (data resulting from simulation tools) in that some experimental processing data (A) serves as input to a process model which in turn outputs some microstructural feature (W), which is stored in the database. This virtual microstructure data is then combined with measured microstructural data (B) and provided as input to a micromechanics and/or statistical mechanics analysis package, which then generates materials property data (X, Y), which again is stored in the database. This property data (X, Y) is then subject to experimental validation (E, F) and also used in some continuum-level analysis package (e.g., finite element analysis (FEA)) to assess some

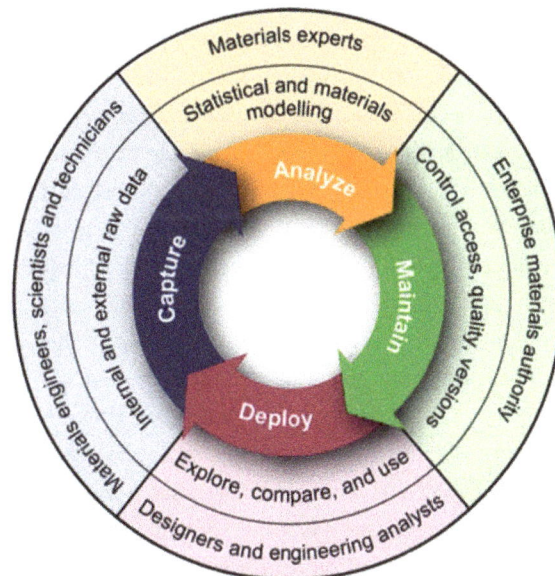

Figure 2 Four aspects of material data life cycle as defined by the MDMC. The Material Data Management Consortium (MDMC) is a group of aerospace and energy sector organizations (both industrial and governmental) that have joined forces to develop best practices and associated software tools to integrate material and structural information technology with the realities of practical product design and advanced research. This group was established in 2002 through collaboration with ASM International, NASA Glenn Research Center, and Granta Design Limited [16] (see www.mdmc.net [17]).

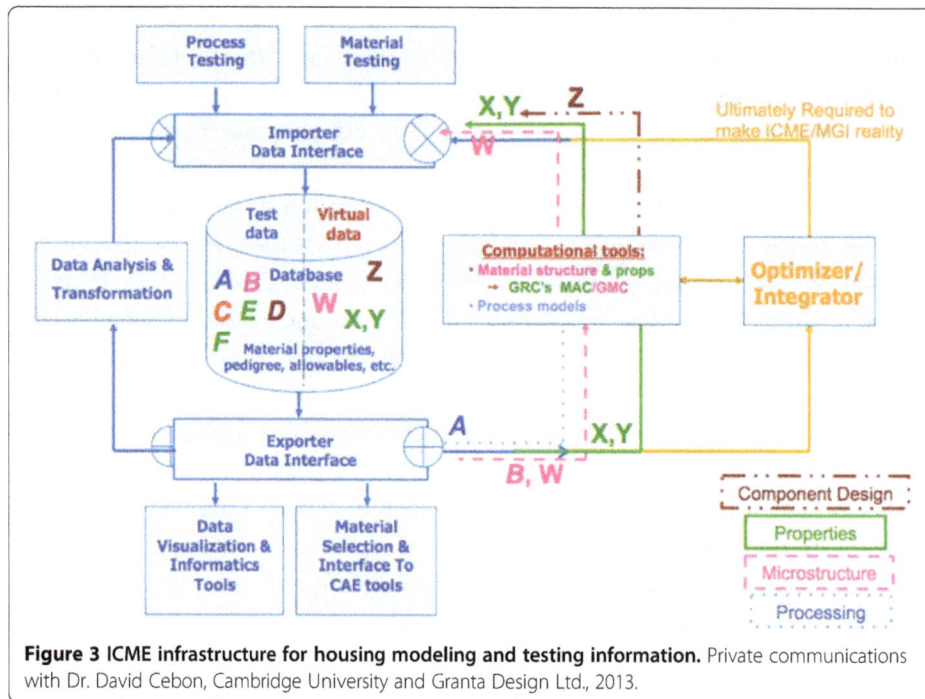

Figure 3 ICME infrastructure for housing modeling and testing information. Private communications with Dr. David Cebon, Cambridge University and Granta Design Ltd., 2013.

performance criteria (e.g., fatigue life, creep rupture, buckling load) Z, which is again stored in the database. Clearly such an information management infrastructure not only enables the capture, analysis, dissemination, and maintenance of various types of data but also facilitates the verification and validation of model output and certification of toolsets at multiple length scales. Also, once all of the input/output protocols are established, it can enable the seamless integration of these toolsets with optimization (e.g., OpenMDAO [7]) algorithms that will provide the final linkage of processing to performance criteria - thus realizing true ICME.

In this paper, our interests lie in identifying the challenges, best practices, and required schema with associated attributes to make the integration of virtual data and test data, described in Figure 3, a reality. Specifically, we will discuss and demonstrate the information management system, based on the Granta MI system, being developed at NASA Glenn Research Center (GRC) for storing not only experimental data (exploratory, characterization, and validation test data, see [4]) but also simulation data (both correlation and predictions) resulting from constitutive modeling activities of both monolithic metals and composite materials. This integration is the first step in our attempt to connect both simulation and experimental data at various scales. Consequently, illustrative emphasis will be placed on the requirements (schema and attributes) for the material/model information management *software*, rather than on the *data* contained within the systems. In the 'Materials information management system' section, the challenges, best practices, and required schema are described, while in the 'Micromechanics of composite materials and structures' section, a micromechanics analysis code and multiscale framework under development at NASA GRC are briefly introduced so that in the 'Linkage of experimental and virtual data via Composite Model Table' section the multiscale linkage between experimental and virtual composite data can be discussed.

Methods

Materials information management system

The Material Data Management Consortium (MDMC) has defined the material data *life cycle* (see Figure 2) in an engineering organization as:

A. Capturing/consolidating materials data;

B. Analyzing materials data;

C. Managing and maintaining the information resource;

D. Deploying and using materials information.

Clearly, this life cycle can be applied similarly to other types of data associated with constitutive models, software tools in general, documentation/reference data, etc. In general, data is captured and consolidated from external sources, legacy databases, as well as internal (possibly proprietary) testing programs. Next, data is analyzed and integrated to create/discover useful information pertinent to the various length scales. The third stage of the data life cycle is the continual maintenance of the whole system (the data and information generated as well as the relationships, or links, between them), with the last, but still crucial, step being the deployment (dissemination) of the right information, to the right people, at the right time, and in the right format. Note that the middle ring of Figure 2 provides additional information regarding the type of data utilized and functions performed during each phase in the data life cycle, while the outermost ring details the individuals most likely responsible for these functions.

To support the various activities throughout the data life cycle, it is preferable to have a single, central source, in which all relevant data is captured and consolidated from 'birth' to 'death' and a variety of software tools are fully integrated (preferably seamless). These tools (as depicted in Figure 3) range from i) data input, ii) reduction/analysis, iii) visualization, iv) reporting, v) process/microstructure/property/performance models (in the case of ICME), vi) material parameter estimation (of both actual and 'virtual' materials), vii) statistical and other analyses to reduce the data to a form usable by designers and analysts - for example, calculation of 'design allowables,' viii) product life cycle management (PLM), and ix) structural analysis codes that utilize a central database. Note that the models and tools listed in (iv) and (v) can operate on a variety of different length scales, thus potentially requiring scale-specific attributes. An example of a micromechanics (microstructure/property/performance) analysis code, known as MAC/GMC, that enables both the 'design of' and 'design with' composite materials will be described and illustrated in the 'Micromechanics of composite materials and structures' section of this paper. If the resulting predicted properties (i.e., virtual data) are stored in the database as well, then it is straightforward to validate such methods and models by direct comparison with actual test data. These tools should enable material and structural engineers to input, manage, and utilize information in an efficient, reliable, and user-friendly way as possible. Finally, these tools should also enable enterprise-wide (even worldwide) solution or access.

Capture

To maximize the impact on the material and structural discipline practitioner and/or researcher, more than just specific predefined (generally accepted) point-wise property values/information needs to be captured from both tests and simulations. In fact, it is essential that a best practice software infrastructure i) has the ability to capture a materials fundamental *multiaxial* response spectrum (under a variety of loading conditions), along with its full pedigree (e.g., chemistry, processing, heat treatment, microstructure, and testing information) for subsequent analysis and modeling; ii) has the ability to capture the application potential of a given material system, be it monolithic, composite, multifunctional, etc.; and iii) enables contextual linkage and association of tacit (or hidden) knowledge (e.g., insight, intuition, skills, experience, and other knowledge that has not been formally shared) within a given organization [3].

Analyzing materials data

For most organizations, a corporate materials database is a dynamic resource - they want to continually add data and to analyze that data to generate new or updated information. This requires software that can process, manipulate, and perform calculations based upon the data. For example, materials experts need software to process raw materials test data and analyze it in order to create approved design data for wider publication. They must update and refine this information and prepare it for use in specialized applications, such as statistical process control or constitutive-life modeling. Such tools are highly specialized and may be developed in-house, come from academic or commercial collaborators, or be purchased. Table 1 lists some examples.

Whatever the exact nature or source of such software tools, best practice materials information management requires that these tools:

- Be able to be used together so that they combine to offer the range of analyses required by materials scientists and engineers - from single test results, to multiple points, to multiple curves;
- Be fully integrated with the information management system, so that data is extracted directly into the analysis tool and results are saved directly back into the correct locations in the database (see the 'Linkage of experimental and virtual data via Composite Model Table' section);
- Their results be permanently linked to raw input data and the details of the analyses performed, so as to maintain full traceability.

Table 1 Examples of analysis tools needed by materials experts

	Examples
Property estimation tools	Thermo-Calc, CALPHAD, MAC/GMC, etc.
Processing test data	Tensile tests, compression tests, creep, fatigue crack growth, E399 fracture toughness, etc.
Material selection/ substitution	CES Selector 2015, MatWeb
Deformation Models	Fit test load/stress, total strain, and/or inelastic strain as a function of time at various constant temperatures (tensile, creep, relaxation, cyclic, step tests, etc.). For example, elastic, viscoelastic, or generalized viscoelastoplastic models. See refs. [18-21]
Damage/Life Models	Stress vs. life curves for stress-controlled cyclic tests using models such as the Basquin model, the Life power model, the Ramberg-Osgood model. Creep strain vs. time, for creep and creep rupture: Larson Miller model or Kachanov type continuum damage mechanics (CDM) model. See refs. [18-21]

Maintaining materials information

Establishing a 'gold source' of materials information is not enough, as this source must also be protected, nurtured, and maintained. A number of data management features are critical to this process: i) traceability, ii) access control, iii) version control, and iv) data quality control as discussed in detail in references [6,8]. Perhaps the most important requirement for best practice materials information management is the ability to trace relevant information on the materials beyond their property data. Knowing a material's pedigree information can help users understand and correctly apply the materials in component designs and constructions. It also provides important information (processing, microstructure, etc.) and references required for improving the materials properties or developing new materials. Most importantly, it is *irrational* to be confident in the use of any data if its pedigree is unknown, as using un-pedigreed data (be it simulation or experimental) involves an extreme risk for safety critical structural applications. While today it is common practice (particularly in research) to use data with questionable pedigrees, it is precisely this background data that is essential to capture, analyze, and maintain if ICME is to become a reality in industrial applications. Consequently, the design of the data schema becomes the major issue in ensuring traceability. Note that, to enable both high traceability and high scalability, separating the individual data categories and connecting them with adequate links becomes an essential attribute of any fit-for-purpose information management system. For example, raw, statistical, and design data are considered to be the core data categories, while pedigree, microstructure, testing, application, in-service environment and exposure, and reference data are normally deemed background information.

NASA GRC's Granta MI® installation, illustrated in Figure 4, is an example of such a fit-for-purpose information management system, wherein NASA GRC's data schema (see Figure 5), an extension of the MDMC data schema, has been specifically designed to enable ICME activities. For example, the *microstructure* information category (table) (see Table 2 for its details of contents) is separated from other material pedigree tables, thus enabling one to go directly to this table and quickly locate typical microstructural images (see Figure 6), and then trace backwards through the links to the other associated material pedigree tables, raw test results, and processing history that produced the specific microstructure. Microscopy information, however, which is associated with changes during testing (due to either mechanical or thermal loading) or subsequent to testing (e.g., failure surface analysis), is typically specimen-specific and thus is stored in the specific specimen record located within the various Test Data tables.

Likewise, to enable scale-specific experimental and model simulation (virtual) data to coexist in the same database, tables associated with 'model pedigree' information (e.g., Deformation, Damage, and Composite) have also been included in the data schema (see Figure 5). Virtual data is an outcome from running some form of model/analysis software tool. For example, in the case of mechanics of materials, this can be as straightforward as exercising a given constitutive model (the simplest being isotropic Hooke's law, which involves only two parameters (e.g., Young's modulus and Poisson's ratio) or as complex as a general nonlinear finite element analysis of a complex structural component resulting in complex response spectrums. In either case, it is essential to understand/record the fundamental assumptions (material system, material anisotropy, linear and nonlinear behavior, boundary conditions, etc.), pertinent model

a) Material tab

b) Model tab (Composite)

Figure 4 NASA GRC's customization of GRANTA MI materials information management system.

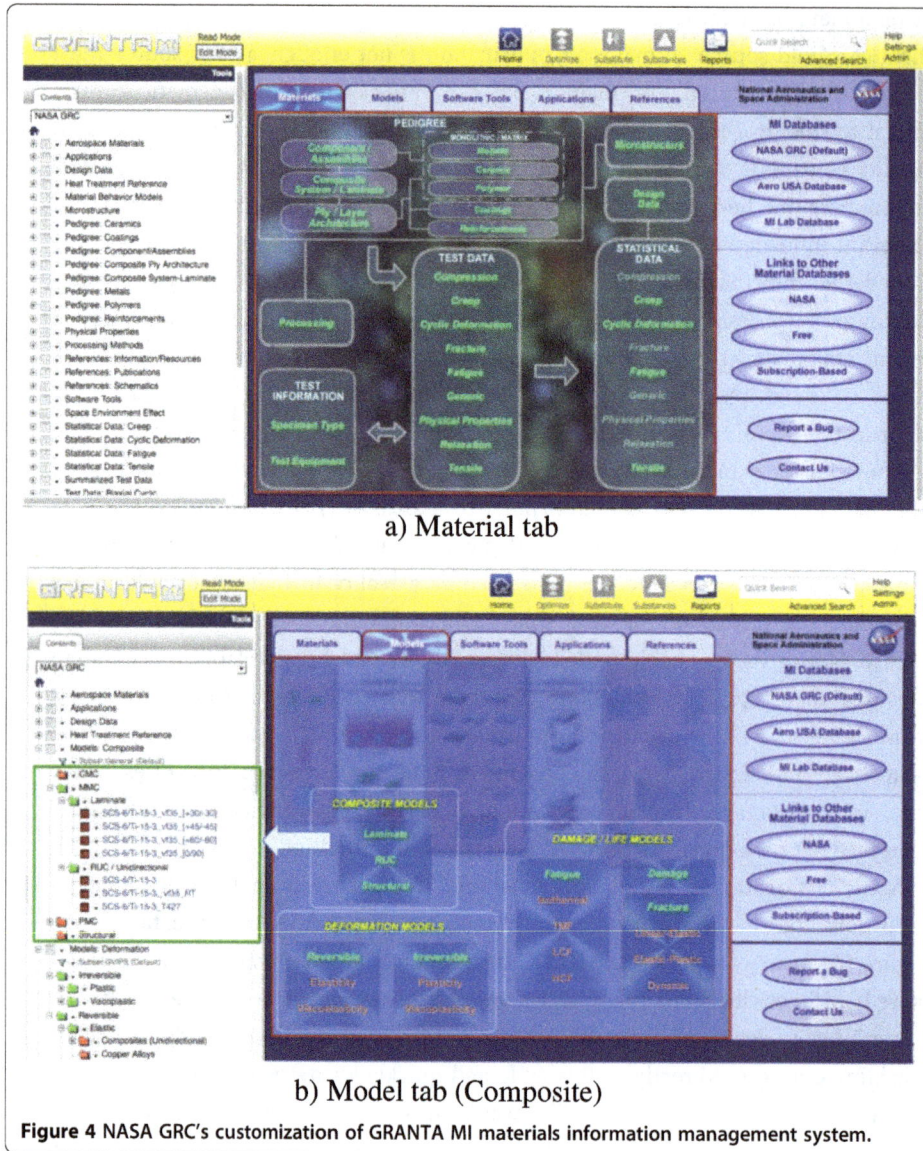

parameters, loading conditions, etc., along with the resulting simulation data itself, in order to properly connect experimental data with simulation data. One might ask, 'Why should I store the resulting simulation data?' The benefits of storing simulation data along with their pedigree information are fourfold: 1) it allows immediate comparison between experiment and simulation, thus enabling an assessment of the accuracy of both the correlation ability and/or predictive ability of the model, 2) it enables periodic re-assessment of the model's accuracy as the experimental data set grows, thus indicating when the model's characterization needs to be updated, 3) it provides future generations with benchmark curves to confirm the version of the model being used or to verify its re-implementation by someone else, and 4) it allows complete traceability, from model version to experimental data used for correlation. Any researcher or analyst who has attempted to reproduce the modeling results of a coworker, or even their own modeling results after a number of years, can attest to the value of storing and tracking simulation data and pedigree.

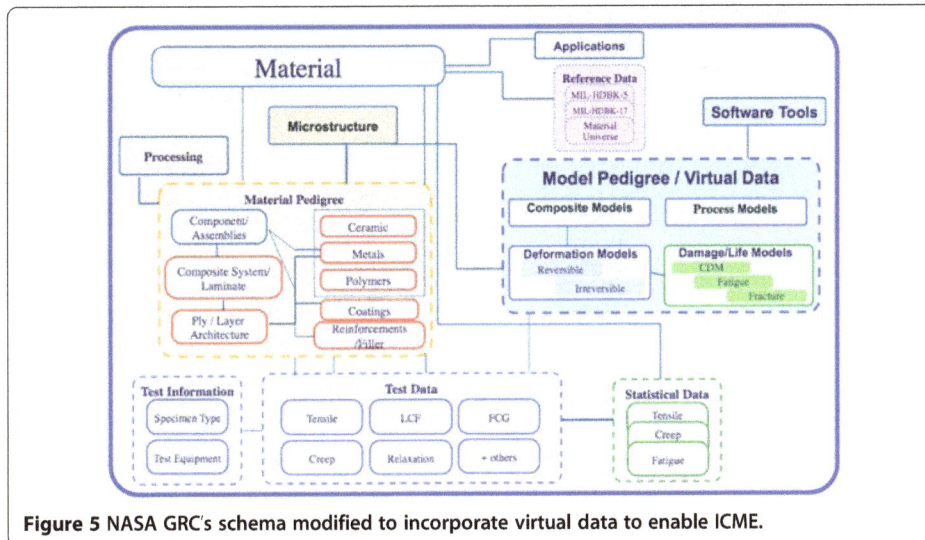

Figure 5 NASA GRC's schema modified to incorporate virtual data to enable ICME.

Obviously, ICME involves a wide variety of models (e.g., process models, internal structure models, and constitutive models) as indicated in Table 1 and thus necessitates a versatile schema. In Arnold et al. [6], the specifics of the schema (i.e., required attributes) and the format (e.g., attribute type and record layout) for best storing such information were discussed in detail for storing information limited to monolithic and composite material coupon level data. In the case of monolithic materials (e.g., fiber and matrix), three tables and their associated attributes were defined to enable the complete data life cycle to be handled; these are the following: Deformation Model Table, Damage/Life Model Table, and Software Tools Table (see Figure 5). Whereas, in the case of composite materials, one must think more broadly as multiple length scales can be involved depending upon the approach taken (i.e., macromechanics or micromechanics) to define the material's 'constitutive model.' Consequently, the additional meso- or macroscale above the constituent scale (e.g., that associated with monolithic material) necessitates the introduction of a fourth table, the Composite Table (discussed in detail in the 'Linkage of experimental and virtual data via Composite Model Table' section). Clearly, extension to other scales (e.g., atomistic, processing, microstructure modeling, structural) may require either the addition of new tables with appropriate attributes to the model pedigree group within Figure 5 (e.g., Process Model Table) or new scale-specific attributes to represent each new scale considered. Clearly, the present schema (with its assembly of model pedigree tables) not only allows model information and model parameters to be stored in a location that is easily accessible by FEA or other analysis codes through some type of interface software (e.g., Materials Gateway®) but also stores any associated simulation data necessary to assist in the evaluation, validation/certification, and utilization of these models.

Micromechanics of composite materials and structures

In its broadest context, a composite is anything comprised of two or more entities with a recognizable interface (i.e., distinct internal boundaries) between them. If these internal boundaries are ignored, continuum mechanics can be used to model composite materials as pseudo-homogenous, anisotropic materials with directionally dependent 'effective,' 'homogenized,' or 'smeared' material properties. Micromechanics, on the

Table 2 Attributes for microstructure description

Attributes	Meta-attributes	
General	Sizes	Phase compositions
Specimen ID	Grain size, measured	Percent
Pedigree ID	Standard deviation, ASTM number	Size
Disk ID	As-large-as grain size, ASTM number	Photomicrographs
Location in disk	Number	Description
Relative quench rate	Major axis: average	Etchant
Relative stress relief time	Major axis: standard deviation	Image magnification
Centroid location	Major axis: range	Image width
Centroid - r	Minor axis: average	Image height
Centroid - z	Minor axis: standard deviation	Date photo taken
Centroid - Θ	Minor axis: range	Photographer
Cutup diagrams	Feret diameter: average	RVE (embedded file)
Cutup diagram pictures	Feret diameter: standard deviation	RVE (link to file)
Microstructure	Feret diameter: range	Distance from centroid - x
Graphic	Aspect ratio: average	Distance from centroid - y
Primary γ', area fraction	Aspect ratio: standard deviation	Distance from centroid - z
Secondary γ', area fraction	Aspect ratio: range	Microscopy technique
Tertiary γ', area fraction	Compactness: average	RVE (representative photo)
Minor phases in matrix	Compactness: standard deviation	Distance from centroid - z
	Compactness: range	
Minor phases at grain boundaries	Shape factor: average	Microscopy technique
Histogram: major axis	Shape factor: standard deviation	RVE (representative photo)
Histogram: minor axis	Shape factor: range	
Histogram: Feret diameter		
Mean grain size, ASTM number		
As-large-as grain size, ASTM number		
Grain size, measured		
Standard deviation, ASTM number		
Histogram: aspect ratio		
Histogram: compactness		
Histogram: shape factor		
Supporting graphics		
Micrographs		
Photomicrographs		
Other phases		
Boundary minor phase composition		
Matrix minor phase composition		

other hand, attempts to account for the internal boundaries within a composite material and capture the effects of the composite's internal arrangement. In micromechanics, the individual materials (typically referred to as constituents or phases) that make up a composite are each treated as continua via continuum mechanics models, with their individual properties and arrangement dictating the overall behavior of the composite material. Over the past two decades, NASA GRC has been developing the ImMAC suite of tools for

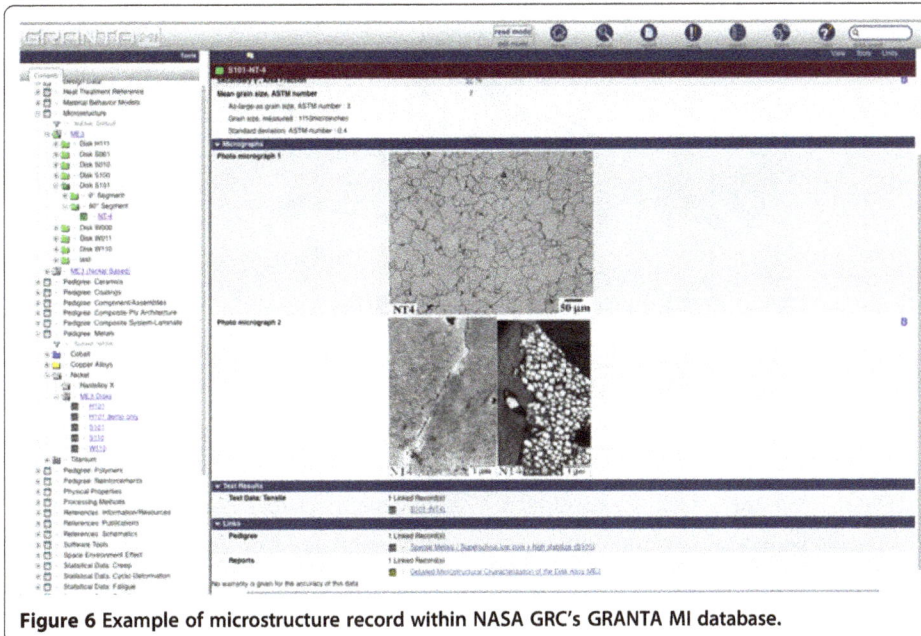

Figure 6 Example of microstructure record within NASA GRC's GRANTA MI database.

analyzing continuous, discontinuous, woven, and smart (piezo-electromagnetic) composite materials and/or structures composed of such materials. MAC/GMC (a comprehensive and versatile stand-alone micromechanics analysis computer code), HyperMAC (the coupling of MAC/GMC micromechanics with the commercial structural sizing software known as HyperSizer [9]), MSGMC (the recursive coupling of micromechanics with micromechanics, for woven composites), and FEAMAC (the coupling of MAC/GMC micromechanics with the commercial finite element code, Abaqus [10]) make up this suite. At the core of these various tools is the well-known method of cells family of micromechanics theories (e.g., method of cells (MOC), generalized method of cells (GMC), and high-fidelity generalized method of cells (HFGMC)) developed by Aboudi and co-workers [11]. These methods provide semi-closed form solutions for determining global anisotropic composite properties in terms of the constituent material properties and arrangement, while also providing the full three-dimensional stresses and strains in each of the constituent subcells. For a detailed, comprehensive discussion on modeling of composite materials, the reader is referred to the book entitled *Micromechanics of Composite Materials: A Generalized Multiscale Analysis Approach* [11]. Micromechanics-based analysis lends itself to ICME in that it links the processing and microstructure of the material directly to the resulting properties and performance of the material/structure, thereby enabling the practitioner to not only 'design with' the material but also concurrently 'design the' material. Consequently, developing a database schema capable of handling a micromechanics approach enables demonstration of an ICME capable (multiscale) framework for composite materials.

The generalized method of cells

It is assumed that a continuously reinforced composite microstructure can be represented as a collection of doubly periodic repeating unit cells (RUCs) containing an arbitrary number of constituents, as shown in Figure 7. The RUC (indicated by a dashed

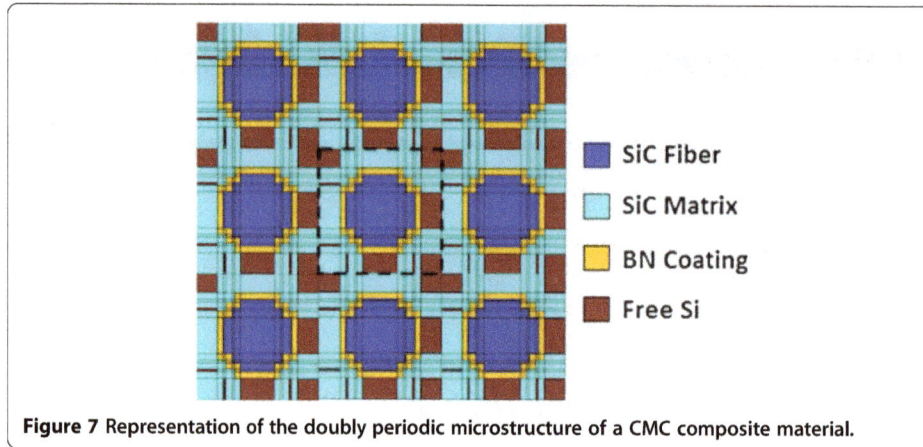

Figure 7 Representation of the doubly periodic microstructure of a CMC composite material.

line in Figure 7) is then discretized into $N_\beta \times N_\gamma$ rectangular subcells (in the case of doubly periodic generalized method of cells (GMC)), as exhibited in Figure 8. Note that triply periodic microstructures (e.g., particulate-reinforced or 3D woven composites), although not discussed here, can also be easily represented as well. Each of these subcells is occupied by one of the constituent materials (e.g., SiC fiber, BN coating, SiC matrix, and free Si in the case of SiC/SiC composites). The number of subcells and the number of materials are completely general. In GMC, a first-order displacement field within the subcells is assumed, and displacement and traction continuity conditions are enforced in an average, integral sense at the subcell interfaces of a discretized RUC. These continuity conditions are used to formulate a set of semi-analytical linear algebraic equations that are solved for the local strains in subcell $(\beta\gamma)$ in terms of globally applied strains or stresses. Then, local constitutive laws can be utilized to obtain the local stresses in subcell $(\beta\gamma)$:

$$\varepsilon^{(\beta\gamma)} = \mathbf{A}^{(\beta\gamma)}\bar{\varepsilon} + \mathbf{D}^{(\beta\gamma)}\left(\varepsilon_s^I + \varepsilon_s^T\right) \tag{1}$$

$$\sigma^{(\beta\gamma)} = \mathbf{C}^{(\beta\gamma)}\left\{\mathbf{A}^{(\beta\gamma)}\bar{\varepsilon} + \mathbf{D}^{(\beta\gamma)}\left(\varepsilon_s^I + \varepsilon_s^T\right) - \left(\varepsilon^{I(\beta\gamma)} + \varepsilon^{T(\beta\gamma)}\right)\right\} \tag{2}$$

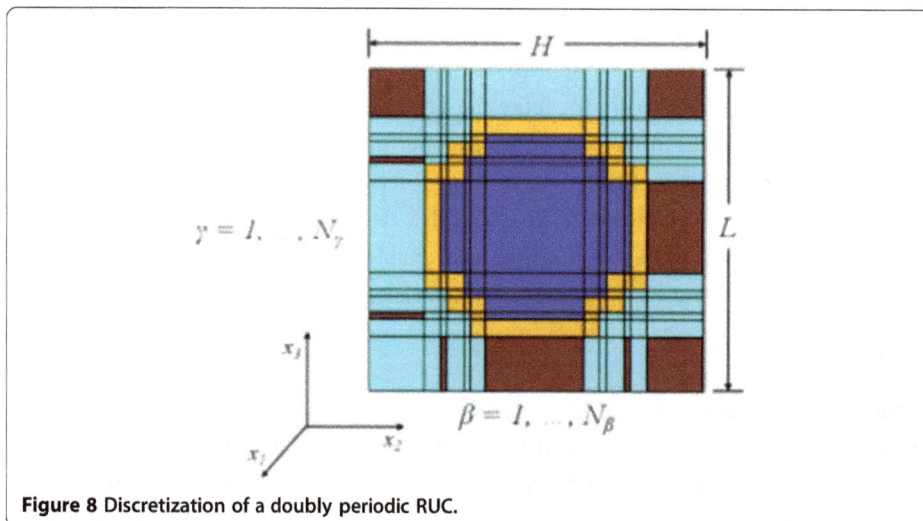

Figure 8 Discretization of a doubly periodic RUC.

where $\boldsymbol{\sigma}$ is the stress tensor, $\mathbf{C}$ is the stiffness tensor, and $\boldsymbol{\varepsilon}$, $\boldsymbol{\varepsilon}^{I}$, and $\boldsymbol{\varepsilon}^{T}$ are the total, inelastic, and thermal strain tensors, respectively, $\boldsymbol{\varepsilon}_{s}^{I}$ and $\boldsymbol{\varepsilon}_{s}^{T}$ are 6 by $N_{\beta}\,N_{\gamma}$ matrices containing all components of the inelastic and thermal strains, respectively, of every subcell (appropriately ordered), $\mathbf{A}^{(\beta\gamma)}$ is the strain concentration tensor, and $\mathbf{D}^{(\beta\gamma)}$ is the thermo-inelastic strain concentration tensor. Then, the generalized constitutive law for the effective, homogenized composite can be formulated as:

$$\bar{\boldsymbol{\sigma}} = \mathbf{C}^{*}\left(\bar{\boldsymbol{\varepsilon}}-\bar{\boldsymbol{\varepsilon}}^{I}-\bar{\boldsymbol{\varepsilon}}^{T}\right) \tag{3}$$

where the effective stiffness tensor, $\mathbf{C}^{*}$, effective inelastic strains, $\bar{\boldsymbol{\varepsilon}}^{I}$, and effective thermal strains, $\bar{\boldsymbol{\varepsilon}}^{T}$, are given by:

$$\mathbf{C}^{*} = \frac{1}{hl}\sum_{\beta=1}^{N_{\beta}}\sum_{\gamma=1}^{N_{\gamma}}h_{\beta}l_{\gamma}\mathbf{C}^{(\beta\gamma)}\mathbf{A}^{(\beta\gamma)} \tag{4}$$

$$\bar{\boldsymbol{\varepsilon}}^{I} = -\frac{\mathbf{C}^{*-1}}{hl}\sum_{\beta=1}^{N_{\beta}}\sum_{\gamma=1}^{N_{\gamma}}h_{\beta}l_{\gamma}\mathbf{C}^{(\beta\gamma)}\left(\mathbf{D}^{(\beta\gamma)}\boldsymbol{\varepsilon}_{s}^{I}-\boldsymbol{\varepsilon}^{I(\beta\gamma)}\right) \tag{5}$$

$$\bar{\boldsymbol{\varepsilon}}^{T} = -\frac{\mathbf{C}^{*-1}}{hl}\sum_{\beta=1}^{N_{\beta}}\sum_{\gamma=1}^{N_{\gamma}}h_{\beta}l_{\gamma}\mathbf{C}^{(\beta\gamma)}\left(\mathbf{D}^{(\beta\gamma)}\boldsymbol{\varepsilon}_{s}^{T}-\boldsymbol{\varepsilon}^{T(\beta\gamma)}\right) \tag{6}$$

h_{β} and l_{γ} are the dimensions of the subcells, h and l are the dimensions of the RUC, and $\bar{\boldsymbol{\sigma}}$ and $\bar{\boldsymbol{\varepsilon}}$ are the effective (homogenized) stress and strain tensors, respectively. Extensive details regarding this derivation can be found in Aboudi et al. [11].

Results and discussion

To illustrate the potential utility of micromechanics for ICME of composite materials, the influence of residual stresses and subsequent post-heat treatment on the laminate response of a $[0°/90°]_{s}$ SiC/SiC CMC composite laminate will be examined. The consituents present within the SiC/SiC RUC include a SiC matrix, SiC fiber, BN coating, and free Si inclusions. Here, the effect of creep of the constituents (wherein it was assumed that the creep of the SiC fiber is less than the creep of the SiC matrix which is less than the creep of the free Si for all temperatures) is accounted for by assuming a simple Norton-Baily power law, $\dot{\varepsilon}^{I} = A\sigma^{n}$, for the fiber, matrix, and free Si inclusions within the matrix. Note that the BN coating is assumed to be elastic, very compliant, and non-damaging in this illustration.

The qualitative effect of including residual stress effects resulting from manufacturing on the proportional limit stress (PLS) and strain to failure (ε_{f}) of a typical simulated tensile test performed at room temperature is shown in Figure 9. Applying a subsequent post-heat treatment (HT) at different temperatures and for different durations shows that the PLS and strain to failure at room temperature can be impacted, as illustrated in Figure 10. Note that the degree of impact (i.e., the amount of redistribution of residual stress) is a function of stress, time, temperature, and microstructure. Clearly, the increase in PLS and decrease in strain to failure, resulting from residual stress as shown in Figure 9, is diminished as the hold time and temperature are increased (see Figure 10). Furthermore, a macromechanics modeling approach could not predict such post-HT tensile behavior, since it is due to internal stress

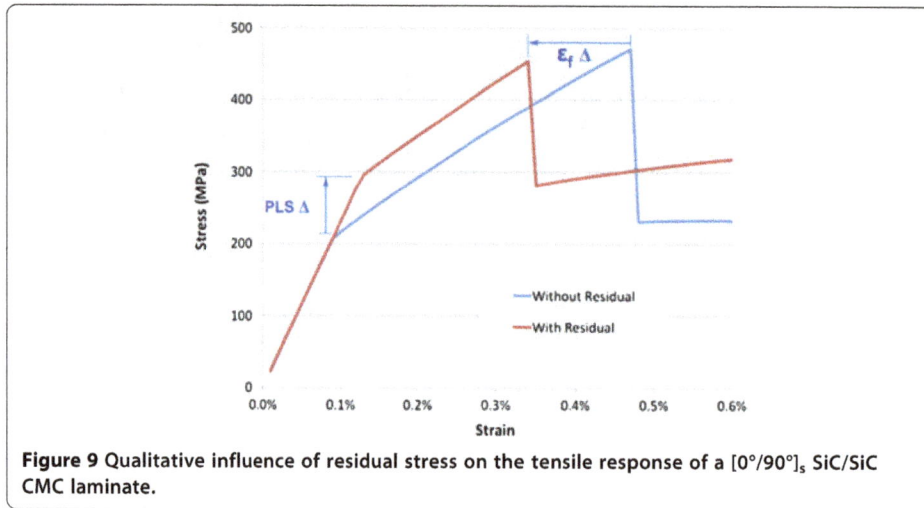

Figure 9 Qualitative influence of residual stress on the tensile response of a [0°/90°]ₛ SiC/SiC CMC laminate.

redistribution within the composite material itself that occurs during (globally) stress-free manufacturing and post-processing conditions. Such behavior has been observed by Bhatt experimentally (private communications). Consequently, micromechanics provides a seamless link between the two non-exclusive viewpoints of designing 'with' the material and designing 'the' material, thereby enabling ICME of composite materials.

For ICME, it is necessary to link the subscale effects to structural performance. As such, a synergistic multiscale framework (which executes concurrent multiscaling in time, but sequential multiscaling in space [12]) has been constructed to simulate the nonlinear response of fiber-reinforced composite structures by modeling the fiber-matrix architecture as an RUC at the microscale using GMC and coupling the microscale to the lamina/laminate level (macroscale) finite element model (FEM). The commercial finite element software, Abaqus [10], is used as the FEM platform, and the MAC/GMC core micromechanics software [13,14] is used to perform microscale calculations. The scales are linked using the FEAMAC software implementation [15], which utilizes various Abaqus/Standard user subroutines. A schematic displaying a

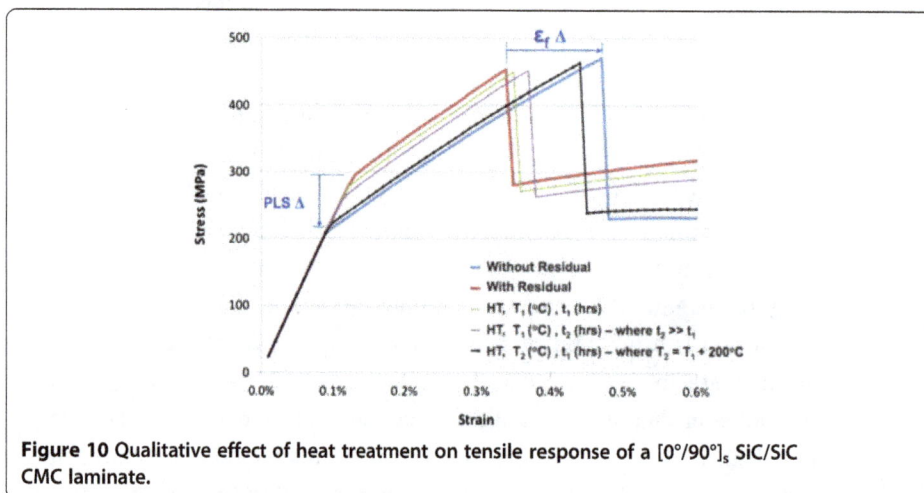

Figure 10 Qualitative effect of heat treatment on tensile response of a [0°/90°]ₛ SiC/SiC CMC laminate.

typical multiscale model using FEAMAC is displayed in Figure 11. The strains at the FEM integration point are applied to the RUC, and the local subcell fields are determined using GMC (this process is referred to as localization). If the subcell material behavior is nonlinear, the local stresses and strains are used to calculate the local stiffnesses, inelastic strains, thermal strains, and/or state variables via the local constituent constitutive laws. Homogenization of the RUC is then performed to obtain the global (effective) stiffnesses, inelastic strains, thermal strains, and/or state variables. The global stresses at the integration point are then calculated using these global, homogenized fields, and the effects of any nonlinear subscale phenomena are introduced into the macroscale through changes in the integration point stress state and stiffness. The global stresses, material Jacobian, and updated state variables at each FEM integration point are then supplied to Abaqus through the user material UMAT subroutine. For complete details on the FEAMAC implementation, the reader is referred to Bednarcyk and Arnold [15] and Aboudi et al. [11].

With the ability to link the GMC micromechanics model, which accounts for processing and microstructure while predicting properties (as discussed in the previous section), with a structural FEM, which simulates performance, the full range of ICME-related scales depicted in Figure 3 has been captured. To briefly illustrate the connection of all aspects of the multiscale (constituent, meso/composite, structural) problems just outlined in Figures 7, 8, 9, 10, and 11 (i.e., material, microstructure, and model pedigree along with test results and software tool description), a repeat of Figure 5, but now with specific names of potential records, is indicated in their pertinent tables within the proposed schema (see Figure 12). Clearly, this is a very high level overview illustration, yet the hope is that it elucidates how one might connect simulation and experimental results and their associated pedigrees together in a single database. Note that the exact location of the simulation results of the structural analysis (i.e., stifffened panel) has yet to be finalized as they could go in a model pedigree table, the application table, or in a PLM system external to the Granta MI database. More specific details regarding layout and associated attributes within the composite and software tables are given in the next section.

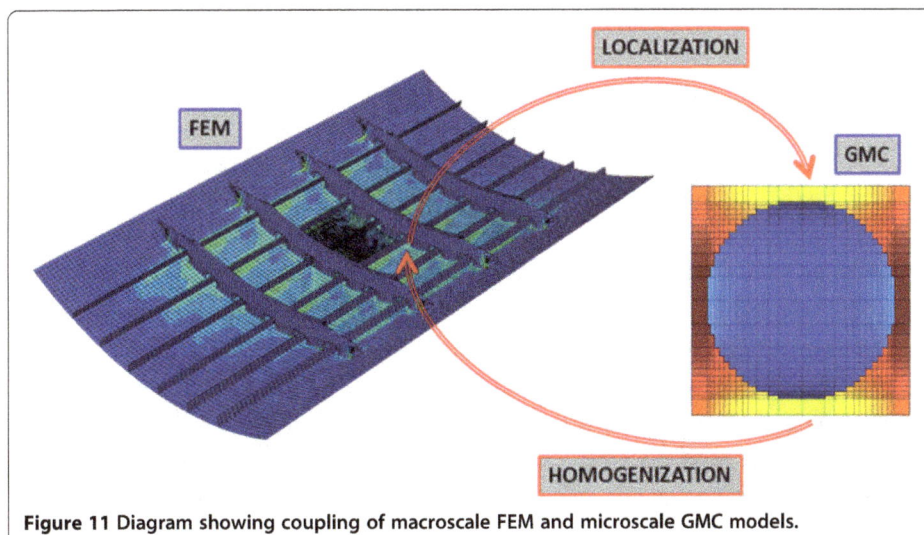

Figure 11 Diagram showing coupling of macroscale FEM and microscale GMC models.

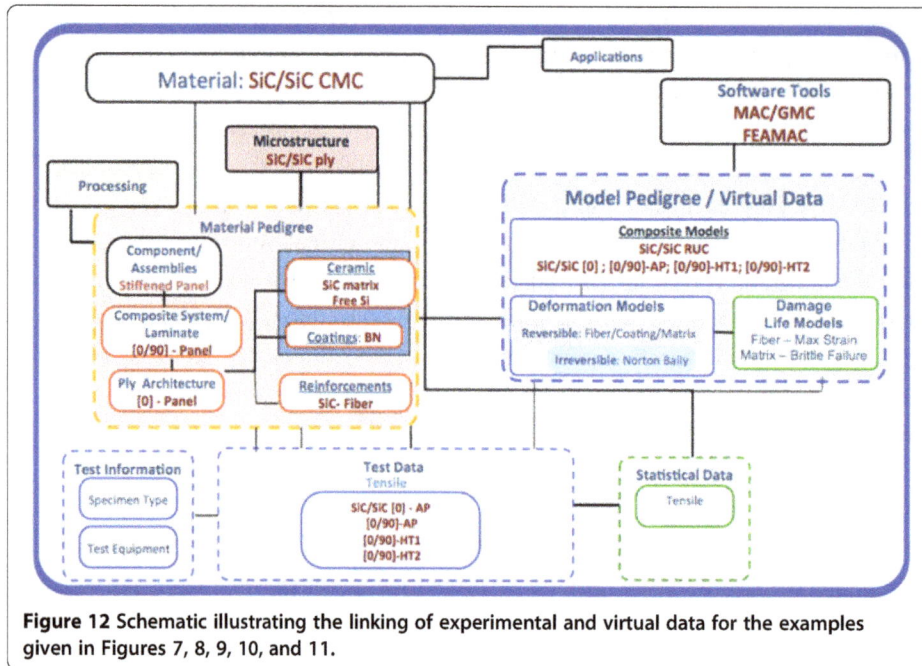

Figure 12 Schematic illustrating the linking of experimental and virtual data for the examples given in Figures 7, 8, 9, 10, and 11.

Linkage of experimental and virtual data via Composite Model Table

As stated previously, foundational to any ICME endeavor (albeit research or product design) is a robust information management system in which both experimental and model simulation (virtual) data coexist, preferably in a single database, at various levels of scale. Just as in the case of experimental data, capturing the pedigree of the material tested is an essential step to enable proper interpretation of results, so too is tracking the pedigree of any simulation (virtual) data entered into the database. Consequently, following the same principles invoked to create material pedigree tables and their linkage to experimental data, a similar thought process was followed to create a set of model pedigree tables that will house all information related to simulations performed at various levels of scale.

The Composite Model Table, as depicted previously in Figure 5, consists of ten basic sections: 'Project Information,' 'Material Description,' 'General Modeling Information,' 'Micromechanics Modeling Information,' 'Laminate Level Modeling Information,' 'Volume Fractions,' 'Multiscale Modeling Information,' 'Composite Test Data Used for Characterization/Validation,' 'Simulation Response,' and 'References' (see Table 3). The first section is associated with the Project Information. The second, Material Description, section, is where the model record is connected to the specific material (or system) that the model is attempting to represent. This is accomplished by linking the material pedigree (via the various attributes in this section (see Table 3) and specifically the material pedigree record link) to the model idealization information contained in the current Model Table record.

The model description section gives the general features of the model, yet in this table, there is no explicit section entitled 'Characterization Information/Parameter Estimation Method' as exists in the Deformation Model Table (described in Arnold et al. [6]). The reason is that this information would be contained in the Deformation Model Table associated with the various constituent materials constitutive models, in the case

Table 3 Layout and attributes for Composite Model Table

Attributes	Type
Project Information	
Performing Organization	STXT
Project Name/Funding Source	STXT
Point of Contact (POC)	STXT
Material Description	
Material	STXT
Material Class	DCT
Commercial Name	STXT
Specific Name	STXT
Material Pedigree Record	Link
Batch Number	STXT
Material Notes	LTXT
General Modeling Information	
Model ID	STXT
Characterization/Analysis Date	DAT
Temperature	PNT
Temperature Range	RNG
Assumptions	LTXT
Micromechanics Modeling Information	
Micromechanics Method	DCT
Micromechanics Analysis Tool	STXT
Micromechanics Tool Information	Link
Micomechanics Input File	Fil
No. of Constituents	INT
RUC/RVE Constitutive Description	TABL
RUC/RVE Image	PIC
Fiber Packing Arrangement	DCT
Effective Thermo-Elastic Composite Properties	TABL
Micromechanics Notes	LTXT
Laminate Level Modeling Information	
Laminate Name	STXT
Laminate Specification	STXT
Architecture Type	DCT
Laminate Pattern	DCT
Laminate Thickness	PNT
Ply Thickness (avg)	PNT
No. of Plies	INT
Laminate Definition	TABL
Laminate Analysis Tool	STXT
Laminate Analysis Tool Information	STXT
Composite Laminate Analysis Input File	FIL
Laminate Notes	LTXT

Table 3 Layout and attributes for Composite Model Table *(Continued)*

Laminate Extensional Stiffness Matrix (A)	TABL
Laminate Coupling Stiffness Matrix (B)	TABL
Laminate Bending Stiffness Matrix (D)	TABL
Volume Fractions	
Total Matrix Volume Fraction	PNT
Total Reinforcement Volume Fraction	PNT
Total Void/Porosity Volume Fraction	PNT
Multiscale Modeling Information	
Multiscale Analysis Tool	DCT
Multiscale Analysis Tool Information	Links
Multiscale Analyses Input File	FIL
Multiscale Modeling Notes	LTXT
Composite Test Data Used for Characterization/Validation	
Tensile Test Data (Linked Records location in layout)	Links
Creep Test Data (Linked Records location in layout)	Links
Relaxation Test Data (Linked Records location in layout)	Links
Cyclic Test Data (Linked Records location in layout)	Links
Generic Test Data (Linked Records location in layout)	Links
Simulation Response	
Stress vs. Strain Response (11 axis)	FDA
Stress vs. Strain Response (22 axis)	FDA
Stress vs. Strain Response (33 axis)	FDA
Stress (11 axis) vs. Time	FDA
Stress (22 axis) vs. Time	FDA
Stress (33 axis) vs. Time	FDA
Total Strain (11 axis) vs. Time	FDA
Total Strain (22 axis) vs. Time	FDA
Total Strain (33 axis) vs. Time	FDA
Shear Stress vs. Shear Strain Response (12 axis)	FDA
Shear Stress vs. Shear Strain Response (13 axis)	FDA
Shear Stress vs. Shear Strain Response (23 axis)	FDA
Shear Stress (12 axis) vs. Time	FDA
Shear Stress (13 axis) vs. Time	FDA
Shear Stress (23 axis) vs. Time	FDA
Total Shear Strain (12 axis) vs. Time	FDA
Total Shear Strain (13 axis) vs. Time	FDA
Total Shear Strain (23 axis) vs. Time	FDA
Force Resultant vs. Midplane normal strain (xx-axis)	FDA
Force Resultant vs. Midplane normal strain (yy-axis)	FDA
Force Resultant vs. Midplane normal strain (xy-axis)	FDA
Moment Resultant vs. Midplane curvature (xx-axis)	FDA
Moment Resultant vs. Midplane curvature (yy-axis)	FDA
Moment Resultant vs. Midplane curvature (xy-axis)	FDA

Table 3 Layout and attributes for Composite Model Table (Continued)

References	
General Modeling Notes	LTXT
Model References	LTXT

DCT, discrete text (specified choices); FDA, functional data attribute (with associated parameters); FIL, allows the association of any file type to a given record; INT, integer value; LOG, logical; LTXT, long text field; PIC, allows association of any image format to a given record; PNT, point value; RNG, range variable; STXT, short text field; TABL, tabular attribute (multiple columns of data - PNT, STXT, DCT, INT, link). Italics are used to assist the reader in locating/connecting specific attributes to discussion in the text and subsequent figures.

of a micromechanics approach. Similarly, in the case of a macromechanics approach, the anisotropic model parameters associated with a given unidirectional 'ply' level material would be stored (along with characterization information (e.g., links to the various tests used to obtain these material parameters)) in their corresponding records in the Deformation Model Table as well.

However, three new sections, specific to composite materials, are present: 'Micromechanics Modeling Information,' 'Laminate Level Modeling Information,' and 'Multiscale Modeling Information,' with only one of these sections per record being populated - depending upon the type of composite analysis being performed. Note that in each of these sections, not only is the analysis tool (again, uniquely defined in the Software Tool Table shown in Table 4) identified but also the associated *input file* required to perform the simulations whose results are stored in the Simulation Response section is required. This is necessary because composites typically require more than just a single set of constitutive model parameters in order to reproduce the simulation results (e.g., in the case of micromechanics, geometric and processing information is also required). Note that the attributes 'Software Tool Used' and 'Regression Software Used' in the Deformation Model Table and 'Micromechanics Analysis Tool,' 'Laminate Analysis Tool,' and 'Multiscale Analysis Tool' in the Composite Model Table allow the best practice of only defining information in one location, yet enabling viewing in multiple locations, to be followed as these attributes link the current model record to the Software Tools Table which contains all the pertinent information regarding the specific model/tool being utilized, i.e., its source code and executable - see Table 4 for the associated attributes and layout.

Two new tabular attributes are defined to represent the RUC or representative volume element (RVE) information and the laminate-level information. Figure 13 illustrates both types of tabular attributes, where each column heading represents a parameter associated with the given tabular attribute. Figure 13a provides an example of a unidirectional, 35% fiber volume fraction, titanium matrix composite (SCS-6/Ti-15-3) represented using the GMC micromechanics approach. Immediately, one sees that two phases are present (fiber and matrix) and that the fiber phase is modeled as an elastic material with its strength being represented by the Curtin fiber breakage model [11]. The matrix phase is modeled as an elastic/plastic material with its fatigue life represented using the anisotropic nonlinear cumulative damage rule - ADEAL [11]. Similarly, the evolving compliant interface (ECI) debond criterion [11] is used between the fiber and matrix phase.

Figure 13b illustrates a fictitious laminate in which a monolithic Ti-15-3 layer is surrounded by a 35% volume fraction, unidirectional SCS-6/Ti-15-3 ply oriented at +45 on the bottom and −45 on the top. Note that the tabular attribute parameter 'Scale' identifies whether a micromechanics approach (indicated by 'RUC') or macromechanics

Table 4 Layout and attributes for Software Tools Table

Attributes	Type
General Description	
Tool Name	STXT
Version	STXT
Description	STXT
Component/System Application	STXT
Tool Scope	DCT
Method	DCT
Software Required to Execute Code	DCT
Other Software Required to Execute Code	STXT
Integration With Other Software	STXT
Website	HYP
Availability	DCT
Last Update (Year)	DCT
Description Notes	LTXT
Classification	STXT
Analysis Design	LOG
Lifing	LOG
Optimization	LOG
Thermal/Heat Transfer	LOG
Thermodynamics	LOG
CFD	LOG
Data Analysis	LOG
Other Classification	STXT
Domain	
Length Scale	DCT
Temporal Scale	DCT
Multiaxiality	DCT
Variables	DCT
Domain Notes	LTXT
Material System Applicability	PNT
Material Independent	LOG
Metallic	LOG
Ceramic	LOG
Polymer	LOG
Composite/Continuous	LOG
Composite/Discontinuous	LOG
Composite/Woven	LOG
Multifunctional	LOG
Smart	LOG
Nano	LOG
Other Material(s)	STXT
Material System Notes	LTXT
Material Description	
Material Directionality	DCT

Table 4 Layout and attributes for Software Tools Table *(Continued)*

Material Scope	DCT
Material Response	DCT
Geometric Description	DCT
Reversible	LOG
Irreversible	LOG
Material Description Notes	LTXT
Platform Supported	
PC	LOG
Mac	LOG
Operating System Supported	
Windows 8	LOG
Windows 7	LOG
Windows NT	LOG
MacOS	LOG
Unix	LOG
Linux	LOG
Operating System Notes	LTXT
Documentation	
Reference Manual	LOG
User's Manual	LOG
References	LTXT
Verification/Validation Method	
Analytical	DCT
Computation	DCT
Experimental	DCT
Verification Notes	LTXT
Software Technology Readiness Level (TRL)	
TRL (1-9)	DCT
Readiness Notes	LTXT
Availability	
Approved for General Release	LOG
Security Classification	DCT
Availability Category	DCT
Sensitivity	DCT
Distribution Limitations	DCT
Limited Until (month/year)	STXT
Point of Contact (POC)	STXT
POC's Organization	STXT
Availability Notes	LTXT
Source Code	
Development Language	DCT
Development Language (other)	STXT
Source Code Available	LOG
Source Code Availability Cat.	DCT
Source Code POC	STXT

Table 4 Layout and attributes for Software Tools Table *(Continued)*

Source Code Location	HYP
Source Code Notes	LTXT
Ownership Rights	
Developer/Performing Org.	STXT
Sponsoring Organization	STXT
Intellectual Property	DCT
Invention Disclosure Filed	LOG
NASA Case No.	STXT
Distribution Category	DCT
Notes	LTXT
Program	
Project Name/Funding Source	STXT
Contract No.	STXT
Grant No.	STXT
Year Initiated	STXT
Software Development Status	DCT
Year Completed/Terminated	STXT
Project Notes	LTXT
Further Information	
Software Reports	Links

DCT, discrete text (specified choices); FDA, functional data attribute (with associated parameters); HYP, hyperlink; INT, integer value; LOG, logical; LTXT, long text field; PNT, point value; RNG, range variable; STXT, short text field; TABL, tabular attribute (multiple columns of data - PNT, STXT, INT, link).

approach (indicated by 'Effective') is being applied to a given layer. In the case of layers 1 and 3, information regarding the modeling of this composite material would be contained in the RUC composite record named 'SCS6/Ti15-3' whose constitutive description is shown in Figure 13a. Therefore, each record referenced can depict a given scale with the interconnection between the constituent scale and the laminate (meso) scale contained within the laminate information tabular attribute.

Next, the 'Simulation Responses' section is where all virtual data is stored. Currently, these functional data attributes (FDAs), e.g., stress vs. strain response (11 axis) and

Phase	Type	Def. Rev.	Def. Irrev.	Damage/Life	Volume Fraction
1	Fiber	SCS-6		Curtin:SCS-6	35
2	Matrix	Ti-15-3E	InPlas:Ti-15-3	ADEAL-Ti15-3	65
	Debond			ECI-SCS6/Ti-15-3	

Note: Type is DCT parameter (Fiber/Fill/Debond/Interface/Matrix)

a) Example of the *RUC/RVE Constitutive Description* attribute (Table 3) filled out

Ply No.	Ply thickness	Ply Angle	Scale	RUC Record	Def. Rev.	Def. Irrev.	Damage/Life	Architecture	Fiber Volume Fraction
1	0.04	-45	RUC	SCS6/Ti15-3				Square	35
2	0.02	0	Effective		Ti-15-3E	InPlas:Ti-15-3	ADEAL-Ti15-3		
3	0.04	45	RUC	SCS6/Ti15-3				Hexagonal	35

b) Example of the *Laminate Definition* attribute (Table 3) filled out

Figure 13 Example of new tabular attributes to describe the composite pedigree. (a) Example of the *RUC/RVE Constitutive Description* attribute (Table 3) filled out. **(b)** Example of the *Laminate Definition* attribute (Table 3) filled out.

in-plane normal loading (force resultant (N) vs. midplane normal strain (ε)) and curvatures (moment resultant (M) vs. midplane curvatures (κ)), have been assigned 13 parameters in all. Ten of these FDA parameters are common to both the Deformation Model Table and Composite Model Table (i.e., specimen ID, test type, loading rate type, loading rate stress magnitude, loading rate strain magnitude, target type, target value, hold duration, simulation classification, and temperature), with four being identified as discrete (i.e., those associated with type and classification), while three are specific to the Composites Model Table - the volume fraction, orientation/layup (this is a short text data type), and resultant specifier (this is a discrete data type, with options: mechanical, inelastic, and thermal). In this way, multiple loading histories can be stored in a *single* attribute that represents a given graphical plotting space, for example, stress-strain, stress-time, strain-time. Obviously, it will be extremely important to establish a process for maintaining consistency of coordinate systems between measured data and simulation data. This is particularly true when one considers composite materials and multiscale modeling. Consequently, a default coordinate system has been established within the NASA GRC GRANTA MI® database.

Lastly, there is 'References' section containing general notes, links to specific applications (which are stored in the Applications Table) and associated reports which are stored in the Reference Table within the NASA GRC GRANTA MI® database. Note that the word 'links' appears in the column associated with type - to indicate that this 'attribute' is merely the name of the link within MI® and not an actual attribute type itself.

Now given the ability to store simulation data (which can be significantly more voluminous than experimental) the next key question for an organization to address is how much of this type of information do they store. Do they mandate that all simulation data be captured or only those attached to a final product. Coming from a research organization, our current thought is to only capture those simulations associated with a final product - in our case, the product is published works. Such a decision is a difficult but extremely important and necessary one to make, as clearly a trade-off exists between the cost of data acquisition, storage, maintenance, and dissemination and the current and future value of the data being collected. This trade is extremely difficult to make *a priori* as one oftentimes does not comprehend the importance of the data/information until after some time has elapsed and the window of opportunity has passed. Consequently, the desire is to collect as much information as possible at the time to avoid being in an 'if only I had …' situation.

Conclusions

ICME is an *integrated* approach to the design of products, and the materials that comprise them, by linking various length scale-specific relationships across the scales, from processing all the way to product performance. A key ingredient is the linkage with manufacturing processes, which produce internal material structures, and in turn influence material properties and allowables, thus enabling tailoring (engineering) of both material and structure to specific industrial applications. As models become more accurate, their complexity tends to increase, and they rely less and less on simplifying assumptions. This complexity drives the need for more data to be measured, predicted, compared, stored, and tracked. Further, the goals of ICME, to link model results and experiments at multiple scales, drives the need for data/metadata and contextual linkage so that knowledge can be both captured and discovered. This underscores the value of a robust dynamic information management system enabling management of changing proprietary data alongside reference data collections, while ensuring consistency, quality, applicability, and full traceability. Often overlooked as a 'mere database', this information management system should be viewed as a 'necessary' or an 'enabling' infrastructural aspect to ICME.

The benefits of a robust information management infrastructure are threefold. Firstly, it enables the capture, analysis, dissemination, and maintenance of various types of data (both experimental and virtual) for all materials with full traceability and security. Secondly, it will facilitate in the verification and validation of model output and certification of toolsets at multiple length scales, and thirdly, the establishment of input/output protocols enables the seamless integration of toolsets with optimization algorithms to provide final linkage of processing to performance criteria - thus making ICME a reality.

In this paper, we have taken the first step in articulating and implementing a robust ICME schema that incorporates 1) microstructure characteristic specifications, 2) material and model pedigree infrastructure for integrating experimental data with virtual data resulting from simulation models being applied at various levels of scales, and 3) attributes identifying the specific software tool(s) utilized. Further, some of the key requirements for best practice in materials informatics (both real and virtual data), were discussed that will enable organizations to effectively respond to the demands of new material and engineering applications and the pressures of operating in a globalized engineering environment. However, many hurdles (e.g., statistics, uncertainty, and optimization) are yet to be overcome, and further challenges (e.g., data quality evaluation and characterization, data error minimization and prevention, organizational and financial challenges) are to be expected, particularly in the area of ICME information management. However, these challenges are likely to be met as materials information management becomes mainstream and as more organizations demonstrate a return on their investment in technology in this area.

Competing interests
The authors declare that they have no competing interests.

Authors' contribution
SMA conceived of schema, associated attributes and drafted manuscript; FAH as database administrator implemented schema; BAB and EJP as MAC/GMC code developers supported the implementation of composite table attributes and associated composite examples. All authors read and approved the final manuscript.

Acknowledgements
The first two authors are grateful to their colleagues in the MDMC for many useful discussions on the key issues addressed in this paper.

References
1. Council NR (2008) Integrated Computational Materials Engineering: a transformational discipline for improved competitiveness and national security. National Academies Press, Washington, DC
2. Marsden W, Cebon D, Cope E (2011) Managing multi-scale material data for access within ICME environments. In: Arnold SM, Wong T (eds) Tools, models, databases, and simulation tools developed and needed to realize the vision of Integrated Computational Materials Engineering. ASM International, Materials Park, OH
3. Arnold SM, Holland FA, Gabb T, Nathal M, Wong T (2013) The coming ICME data tsunami and what can be done. 54th AIAA/ASME/ASCE/AHS/ACS structures, structural dynamics, materials conference, Boston, MA., 23–27 Apr 2013
4. Arnold SM (2006) Paradigm shift in data content and informatics infrastructure required for generalized constitutive modeling of materials behavior. MRS Bull 31(12):1013–1021
5. Cebon D, Ashby MF (2006) Engineering materials informatics. MRS Bull 31(12):1004–1012

6. Arnold SM, Holland FA, Bednarcyk BA (2014) Robust informatics infrastructure required for ICME: combining virtual and experimental data. In: 55th AIAA/ASMe/ASCE/AHS/SC structures, structural dynamics, and materials conference. National Harbor, Maryland, pp 13–17, Jan 2014, AIAA-2014-04

7. Open MDAO. http://openmdao.org, 01/15/15.

8. Ren W, Cebon D, Arnold SM (2009) Effective materials property information management for the 21st century. In: Proceedings of PVP2009, 2009 ASME pressure vessels and piping division conference. Czech Republic, Prague, pp 26–30, July 2009, PVP2009-77314, pp. 1–10

9. Collier Research Corporation makers of HyperSizer. www.HyperSizer.com, Accessed 15 January 2015.

10. Simulia Abaqus, a Dassault Systemes subsidiary. http://www.3ds.com/products-services/simulia/products/abaqus/. Accessed 15 January 2015.

11. Aboudi J, Arnold SM, Bednarcyk BA (2013) Micromechanics of composite materials: a generalized multiscale analysis approach. Elsevier, Oxford, UK

12. Sullivan RW, Arnold SM (2011) An annotative review of multiscale modeling and its application to scales inherent in the field of ICME. In: Arnold SM, Wong TT (eds) Models, databases, and simulation tools needed for the realization of Integrated Computational Material Engineering. ASM International, Materials Park, OH, pp 6–23

13. Bednarcyk BA, Arnold SM. MAC/GMC 4.0 user's manual – keywords manual. NASA/TM-2002-212077/VOL2, 2002a, NASA Glenn Research Center, Cleveland, OH, USA.

14. Bednarcyk BA, Arnold SM. MAC/GMC 4.0 user's manual – example problems manual. NASA/TM-2002-212077/VOL3, 2002b, NASA Glenn Research Center, Cleveland, OH, USA.

15. Bednarcyk BA, Arnold SM (2006) A framework for performing multiscale stochastic progressive failure analysis of composite structures. In: Proceedings of the 2006 Abaqus user's conference, 23–25 May 2006. MA, Boston

16. Granta Design Limited. http://www.grantadesign.com/. Accessed 19 Mar 2013.

17. Official website of Materials Data Management Consortium. http://mdmc.net. Accessed 19 March 2013.

18. Dowling NE (1999) Mechanical behavior of materials: engineering methods for deformation, fracture, and fatigue. Prentice Hall, New Jersey

19. Lemaitre J, Chaboche JL (1990) Mechanics of solid materials. Cambridge University Press, Cambridge, UK

20. Skrzypek J, Hetnarski R (2000) Plasticity and creep, theory, examples, and problems. CRC Press, Boca Raton, FL, USA

21. Lemaitre J (2001) Handbook of materials behavior models. Academic Press, San Diego, USA

10

Role of cyberinfrastructure in educating the next generation of computational materials scientists

Susan B Sinnott[*] and Simon R Phillpot

* Correspondence: ssinn@mse.ufl.edu
Department of Materials Science and Engineering, University of Florida, Gainesville, FL 32611-6400, USA

Abstract

An overview of cyberinfrastructures developed to advance the field of materials modeling is presented. The role of cyberinfrastructures in educating the next generation of the workforce is also discussed, with an emphasis on the Cyberinfrastructure for Atomistic Simulation (CAMS). The paper concludes with a summary regarding the future outlook of cyberinfrastructures, especially with regard to education.

Keywords: Cyberinfrastructure; Material modeling; Education

Background

The last decade has witnessed tremendous growth in the development and utilization of cyberinfrastructures in a variety of science and engineering disciplines. This includes the materials modeling community, where cyberinfrastructures have been used to enable the sharing of information to advance a specific objective within a team of researchers, or within a specific community to distribute information or establish best practices. For example, the Atomic-scale Friction Research and Education Synergy Hub (AFRESH) [1] was developed to share links to computational tools, data provided by users and mined from the literature, and best practices for modeling friction with atomic-scale simulations.

Cyberinfrastructures have additionally been used to host databases generated from a particular experimental apparatus, from a computational program, and/or from published literature. Having extensive, specialized data gathered in a single location and available to other researchers is extraordinarily beneficial for subsequent material design and discovery. This has therefore been a common way in which cyberinfrastructure has been used, at least in part. For instance, the Materials Project at MIT [2] provides access to databases of material properties generated using first-principles calculations. The focus is primarily on materials for energy storage applications, although the information may be useful for a variety of applications. Another example is the AFLOWLIB [3] cyberinfrastructure, hosted at Duke University, that provides access to an extensive database of metal alloy properties and phase diagrams. AFLOWLIB also hosts databases of material properties that include structural, electronic, and thermoelectric characteristics, to name just a few.

An important use of cyberinfrastructures for computational materials science has been to give broad access to computational tools of various types. For example, in the case of AFRESH access to tools for displaying the output of atomic-scale simulations and examples for their optimal use were provided. Additionally, both the Materials Project and AFLOWLIB allow users to carry electronic structure calculations through the cyberinfrastructure itself at computers located at the hosting institution. Cyberinfrastructures may further serve to organize and disseminate information that is critical for the optimal use of a computational method. This is exemplified by the openKIM project [4] and the NIST Interatomic Potentials Repository Project [5] that catalog interatomic potentials for classical simulations in a unique way so that they may be clearly identified when they are used. These cyberinfrastructures further host their organizational systems in a user friendly way and propagate information to the community as a whole.

An additional important application of cyberinfrastructure in computational materials science and engineering is to enhance the education of the next generation of computational materials scientists. For example, the openKIM project has held several workshops, primarily students and postdocs, about the openKIM cataloging system, including information on how to migrate potentials to the cyberinfrastructure, and to how to fully used the of the openKIM system for potential development.

This educational role is critical and will deepen the understanding of newcomers to the field, assist them with learning to use important technical software for computing and data analysis, and provide a forum for them to share information. Educating the next generation in this manner is also a critical step to ensure the long-term viability of the cyberinfrastructure itself. Indeed, it will ultimately be this next generation who will use, maintain, and expand these platforms.

Review

The Cyberinfrastructure for Atomistic Materials Science (CAMS) [6] is a pilot-project stage platform focused on modeling materials at the atomic scale, with an emphasis on microstructural features. It fills a number of the roles discussed above. For example, it houses an actively curated "virtual library" of microstructure samples, where each sample contains the atomic-scale coordinates of a specified material microstructure. The microstructures that are featured include multiple grain boundaries, surfaces, dislocations, nanocrystalline samples, and related structures. CAMS also links to similar databases to assist the user in finding the structure of interest, reduce duplication across such libraries, and ultimately to encourage the sharing of information across platforms. Users are also encouraged to contribute samples to the CMAS library.

Importantly, CAMS is actively engaged in educational activities (see Figure 1). Part of the motivation for this is to increase use of the virtual library and to grow the number of samples that are contained within it. However, the main motivation is tied to the central mission of CAMS, which is to assist the materials community by serving as a hub for disseminating information on atomic-scale modeling and related methods, especially as they apply to microstructural features within materials. A key target constituency is the next generation of materials modelers, including students, postdocs, junior faculty, and other newcomers to the field.

To reach this audience, in spring 2013 CAMS sponsored and hosted a summer school on "Simulation of Complex Microstructure in Materials" at the University of

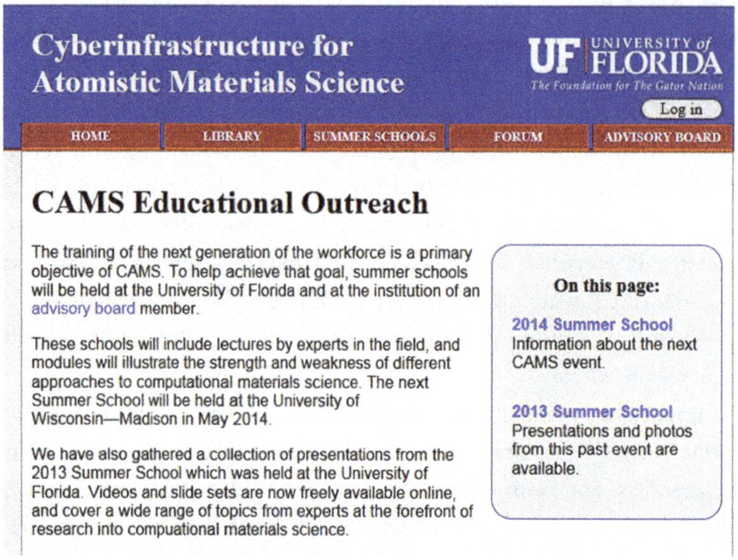

Figure 1 Information page for CAMS summer school.

Florida. More than 50 participants from 20 different U.S. institutions and one German university participated. The instructors for the week-long school were computational and experimental researchers with considerable expertise from multiple U.S. academic institutions, one German university, and U.S. National Laboratories. After each presentation, participants were charged in working in groups to develop questions for the speakers. In so doing, the group members answered many of each other's questions and reinforced the learning that had taken place in the presentation. The lecturer then returned for an extended question and answer session; the questions that the groups then brought forward were sophisticated and well thought out, and thus initiated extensive and detailed further discussions. Many participants, both students and lecturers, commented on the effectiveness of this approach. In addition, the participants were given a training session on how to use and upload structures to the CAMS virtual library; they were further encouraged to contribute structures to it during and after the school. The participants presented research results from their home institution in a poster session and worked together in teams to develop research proposals based on the topics presented at the school. Awards were given to the best posters and best proposals by a panel of judges populated by a diverse mix of instructors. Lastly, the school set aside time for professional development activities, including a discussion led by a representative of the University of Florida's College of Engineering Entrepreneurship Institute, and discussions of their professional lives by a faculty member at a research-intensive institution and a staff member at a US Department of Energy National Laboratory.

The CAMS-run summer school thus contained many opportunities for technical, professional, and interpersonal development on the part of the participants. While travel grants to the participants ensured broad participation, the number of attendees was still limited by the inevitable logistical limitations of any such activity. Therefore, copies of most of the presentations and videos of most of the lectures are provided free of charge on CAMS (see Figure 2) to anyone who goes through a simple registration process. In this way, the impact of the educational effort is disseminated broadly thus further increasing accessibility.

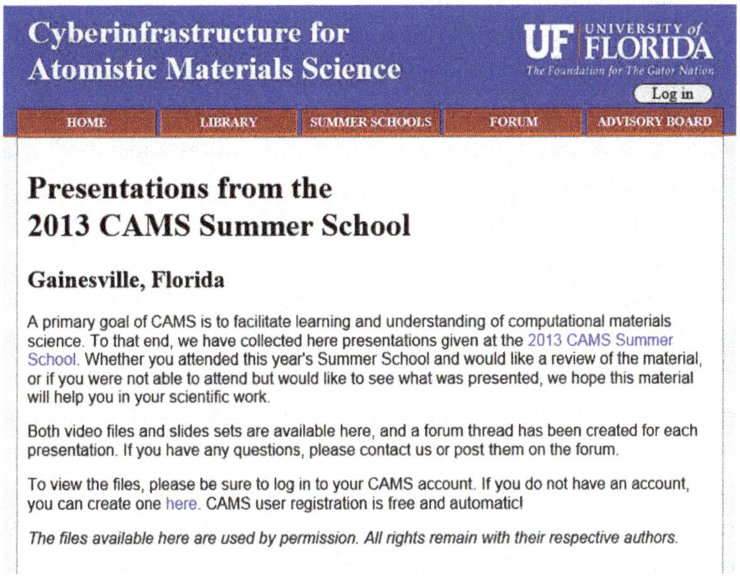

Figure 2 Posted presentations for the 2013 CAMS summer school.

In this respect CAMS joins other, more established cyberinfrastructures that are focused on educational activities. For instance, the CAVS CyberDesign effort at Mississippi State University [7] is a comprehensive wiki site devoted to the topic of integrated computational material engineering. Specifically, this cyberinfrastructure is focused on providing access and guidance to computational materials science methods to enable their use in conjunction with design and manufacturing. Several different models are provided related to materials design and educational courses at a variety of educational institutions, highlighting the power of a cyberinfrastructure to enable the sharing of expertise and educational tools among educators. Similarly, the nanoHUB [8] based at Purdue University contains a large number of computational tools, lectures, and other resources in the area of nanoscience and nonotechnology, with a particular strength in the area of electronic devices.

Conclusions

The future of cyberinfrastructures relies on their ability to successfully meet the needs of their constituencies and to provide resources that are deemed to be valuable. Additionally, resources to sustain their maintenance and continued growth have to be available. The ability of cyberinfrastructures to utilize each other's resources is particularly key, as this will allow each one to magnify the effect of the individual effort. It is also important that they work with other, perhaps less formal, efforts in community building. Efforts such as the Materials Innovation @TMS website (http://materialsinnovation.tms.org/) greatly assist cyberinfrastructures to connect with one another and achieve this goal.

An example of such cooperation is the spring 2014 summer school titled "Transformational Technologies in Molecular Simulations" that will be held at the University of Wisconsin-Madison (UW). This summer school is co-sponsored by CAMS and the MaterialsHUB [9], a cyberinfrastructure focused on the development of new theories and computational tools for the rapid calculation of material properties at the atomic scale. The UW Materials Research Science and Engineering Center (MRSEC) Interdisciplinary Computational Group is also co-sponsoring the summer school. This collaborative interaction promises a wider breadth of topics within the common focus area of interest to these cyberinfrastructures and interdisciplinary group. Ultimately, the collaboration should thus enrich the educational experience of the next generation computational materials scientists that participate in the school or access the subsequent electronically disseminated content.

The common initiatives among the cyberinfrastructures described here to catalog, curate, and distribute content from developers, users, and researchers within the field in such a way that each contributor to the cyberinfrastructure receives full credit is an important advance that, if adopted by the next generation of material modelers, will lead to a number of benefits. These include a new avenue to disseminate and to develop computational tools, to clarify the literature for everyone but especially to newcomers, and to ultimately lower the barrier for entry into the field of computational materials science by the non-expert.

The importance of materials modeling cyberinfrastructures is only expected to increase as the use of mobile electronic devices become more widespread, computational methodologies become a common approach in the material structure–property relationship toolbox, and the next generation of the workforce uses these platforms as an expected part of the modern research endeavor.

Competing interests
The authors declare that they have no competing interests.

Authors' contributions
SBS and SRP both contributed to setting up CAMS and writing the present review article. Both authors read and approved the final manuscript. Both authors did indeed read and approve the final manuscript.

Acknowledgements
The authors gratefully acknowledge the support of the National Science Foundation (DMR-1246173).

References
1. Sinnott SB, Fortes JAB, Bucholz EW, Matsunaga AM (2009) Atomic-scale Friction Research and Education Synergy. In: NSF Engineering Research and Innovation Conference. Honolulu, Hawaii
2. Jain A, Ong P, Hautier G, Chen W, Richards W, Dacek S, Cholia S, Gunter D, Skinner D, Ceder G (2013) The materials project: a materials genome approach to accelerating materials innovation. App Phys Lett Materials 1:011002
3. Curtarolo S, Setyawan W, Wang S, Xue J, Yang K, Taylor R, Nelson L, Hart G, Sanvito S, Buongiorno-Nardelli M, Mingo N, Levy O (2012) AFLOWLIB.ORG: A distributed materials properties repository from high-throughput Ab Initio calculations. Comput Mater Sci 58:227–235
4. Tadmor EB, Elliott RS, Sethna JP, Miller RE, Becker CA (2011) Knowledgebase of Interatomic Models (KIM). Available via https://openkim.org
5. Becker CA, Tavazza F, Trautt ZT, de Macedo RA B (2013) Considerations for choosing and using force fields and interatomic potentials in materials science and engineering. Curr Opinion Solid State Mater Sci 17:277–283, Available via http://www.ctcms.nist.gov/potentials/
6. (2013) Cyberinfrastructure for Atomistic Materials Science (CAMS). Available via http://cams.mse.ufl.edu
7. (2013) CAVS Engineering Virtual Organization for CyberDesign. Available via https://icme.hpc.msstate.edu/mediawiki/index.php/Main_Page
8. Madhavan K, Zentner L, Farnsworth V, Shivarajapura S, Zentner M, Denny N, Klimeck G (2013) nanoHUB.org: Cloud-based services for nanoscale modeling, simulation, and education. Nanotechnology Reviews 2:107–117, Available via https://nanoHUB.org
9. (2013) MaterialsHUB. Available via https://materialshub.org

Permissions

List of Contributors

Yang Jiao
Materials Science and Engineering, Arizona State University, Tempe, AZ 85287-6206, USA

Nikhilesh Chawla
Materials Science and Engineering, Arizona State University, Tempe, AZ 85287-6206, USA

Ann Bolcavage
Rolls-Royce Corporation, Indianapolis, IN 46225, USA

Paul D Brown
Rolls-Royce PLC, Derby DE24 8BJ, UK

Robert Cedoz
Rolls-Royce Corporation, Indianapolis, IN 46225, USA

Nate Cooper
Rolls-Royce Corporation, Indianapolis, IN 46225, USA

Chris Deaton
Rolls-Royce Corporation, Indianapolis, IN 46225, USA

Daniel R Hartman
Rolls-Royce Corporation, Indianapolis, IN 46225, USA

Akin Keskin
Rolls-Royce PLC, Derby DE24 8BJ, UK

Kong Ma
Rolls-Royce Corporation, Indianapolis, IN 46225, USA

John F Matlik
Rolls-Royce Corporation, Indianapolis, IN 46225, USA

Girish Modgil
Rolls-Royce Corporation, Indianapolis, IN 46225, USA

Jeffrey D Stillinger
Rolls-Royce Corporation, Indianapolis, IN 46225, USA

Ankit Agrawal
Department of Electrical Engineering and Computer Science, Northwestern University, Evanston, IL, USA

Parijat D Deshpande
Tata Research Development and Design Centre, Tata Consultancy Services, Pune, Maharashtra, India

Ahmet Cecen
School of Computational Science and Engineering, Georgia Institute of Technology, Atlanta, GA, USA

Gautham P Basavarsu
Tata Research Development and Design Centre, Tata Consultancy Services, Pune, Maharashtra, India

Alok N Choudhary
Department of Electrical Engineering and Computer Science, Northwestern University, Evanston, IL, USA

Surya R Kalidindi
School of Computational Science and Engineering, Georgia Institute of Technology, Atlanta, GA, USA
Woodruff School of Mechanical Engineering, Georgia Institute of Technology, Atlanta, GA, USA

Bo Sundman
INSTN, CEA Saclay, 91191 Saclay, France

Ursula R Kattner
Department of Materials Science and Engineering Division, National Institute of Standards and Technology, 100 Bureau Dr., Gaithersburg, MD 20899, USA

Mauro Palumbo
ICAMS, Ruhr University Bochum, Universitätsstr. 150, 44780 Bochum, Germany

Suzana G Fries
ICAMS, Ruhr University Bochum, Universitätsstr. 150, 44780 Bochum, Germany

Zi-Kui Liu
Department of Materials Science and Engineering Division, National Institute of Standards and Technology, 100 Bureau Dr., Gaithersburg, MD 20899, USA

David L McDowell
ICAMS, Ruhr University Bochum, Universitätsstr. 150, 44780 Bochum, Germany

Royan J D'Mello
Composite Structures Laboratory, Department of Aerospace Engineering, University of Michigan, 1320 Beal Avenue, Ann Arbor, MI 48109-2140, USA
William E. Boeing Department of Aeronautics and Astronautics, University of Washington, Seattle, WA 98195-2400, USA

Marianna Maiarù
Composite Structures Laboratory, Department of Aerospace Engineering, University of Michigan, 1320 Beal Avenue, Ann Arbor, MI 48109-2140, USA
William E. Boeing Department of Aeronautics and Astronautics, University of Washington, Seattle, WA 98195-2400, USA

Anthony M Waas
Composite Structures Laboratory, Department of Aerospace Engineering, University of Michigan, 1320 Beal Avenue, Ann Arbor, MI 48109-2140, USA
William E. Boeing Department of Aeronautics and Astronautics, University of Washington, Seattle, WA 98195-2400, USA

Lawrence E Pado
The Boeing Company, 8900 Frost Ave., Bldg 245 MC S245-1260, 63134 Berkeley, MO, USA

Steven M Arnold
NASA Glenn Research Center, 21000 Brookpark Road, Cleveland, OH 44135, USA

Frederic A Holland Jr
NASA Glenn Research Center, 21000 Brookpark Road, Cleveland, OH 44135, USA

Brett A Bednarcyk
NASA Glenn Research Center, 21000 Brookpark Road, Cleveland, OH 44135, USA

Evan J Pineda
NASA Glenn Research Center, 21000 Brookpark Road, Cleveland, OH 44135, USA

Susan B Sinnott
Department of Materials Science and Engineering, University of Florida, Gainesville, FL 32611-6400, USA

Simon R Phillpot
Department of Materials Science and Engineering, University of Florida, Gainesville, FL 32611-6400, USA